CHEMISTRY
and
MICROSTRUCTURE
of
SOLIDIFIED WASTE
FORMS

Edited by

Roger D. Spence
Oak Ridge National Laboratory
Oak Ridge, Tennessee

LEWIS PUBLISHERS
Boca Raton   Ann Arbor   London   Tokyo

Library of Congress Cataloging-in-Publication Data

Chemistry and microstructure of solidified waste forms/
    edited by Roger D. Spence.
       p. cm.
"Contains proceedings from the Chemistry and Microstructure of Solidified
Waste Forms Symposium sponsored by the Environmental Division of the
American Chemical Society (ACS) at the ACS National Meeting in New York
City on August 25–30, 1991"—Pref.
    Includes bibliographical references and index.
    ISBN 0-87371-748-1
    1. Hazardous wastes—Analysis—Congresses. 2. Cement composites—
Analysis—Congresses. I. Spence, R. D. (Roger David), 1948–
TD1030.C48   1992
628.4′2—dc20   92-36580
                                                            CIP

Direct all inquiries to CRC Press, Inc., 2000 Corporate Blvd., N.W.,
Boca Raton, Florida 33431.

PRINTED IN THE UNITED STATES OF AMERICA
1 2 3 4 5 6 7 8 9 0
Printed on acid-free paper

PREFACE

This volume contains proceedings from the *Chemistry and Microstructure of Solidified Waste Forms* Symposium sponsored by the Environmental Division of the American Chemical Society (ACS) at the ACS National Meeting in New York City on August 25 to 30, 1991. These published proceedings contain written versions of most, but not all, of the papers presented orally in New York. The symposium and its proceedings were limited to cementitious, or cement-based, solidified/stabilized waste forms. The bulk of solidified/stabilized wastes in both radioactive waste disposal and hazardous waste disposal consist of cementitious waste forms. Materials other than cement are used to generate cementitious waste forms, including lime, fly ash, ground granulated blast furnace slag, and combinations. Cementitious waste forms do not include waste encapsulated in polymers, such as bitumen, polyethylene, and vinyl ester styrene, or vitrified in glass waste forms.

Despite common usage worldwide, the fundamental chemistry and microstructure of these waste forms is not well known, although some fundamental knowledge does exist. Groups scattered around the world have been studying how contaminants are immobilized in these waste forms and the effect of these contaminants on the chemistry and microstructure of the cement binders, for years, if not decades. The purpose of this symposium, and subsequently this book, was to tap into these centers of knowledge and learn what these centers had learned over the years. Leaching as a main topic was purposely avoided because the large amount of work done on leaching requires a separate symposium and volume, or volumes. Leaching was not totally avoided; after all, the main purpose of understanding the chemistry and microstructure is to understand how contaminants are immobilized in and how they are released from these waste forms, i.e., leached. Therefore, leaching is addressed insofar as attempting to bridge the gap between fundamental understanding of mechanisms and observation of performance.

Many, but not all, of these centers of knowledge were contacted and asked to make a contribution on the title subject. Not surprisingly, not all could respond, although interest was great in such an enterprise. Unfortunately, with the exception of Prof. Glasser from the University of Aberdeen, the only respondents were from North America, limiting the international scope. Nevertheless, there is a wealth of experience and knowledge presented by authors who could respond.

It will become apparent that not all fundamental mechanisms involved in solidifying waste into cementitious waste forms are known yet, and much remains to be done before one can design a waste form for a given waste, from fundamental principles, or predict performance over decades and centuries with confidence. This is hardly surprising, since the fundamental mechanisms of the chemistry and microstructure of cement pastes are still not completely known with confidence, and they have been studied much longer than waste forms. Unlike the case with waste forms, many books have been published about cement pastes. For example, Prof. H. F. W. Taylor from the University of Aberdeen, one of the most distinguished cement chemists worldwide, recently published a book on cement chemistry, *Cement Chemistry,* published in 1990 by Academic Press. Jesse Conner, a leader in the field of solidification/ stabilization for decades, recently published a book on cementitious waste forms, *Chemical Fixation and Solidification of Hazardous Wastes,* published in 1990 by Van Nostrand Reinhold. Cements and concretes are complex heterogeneous matrices. The fate of waste contaminants in this complex environment is not easy to determine scientifically. It is much easier to engineer a waste form and test some overall property such as leach resistance than to prove the fundamental mechanisms of immobilization and leaching. This approach essentially treats the waste form as a "black box". Unexpected behavior and surprises can result from such an approach because of a lack of true understanding of the mechanisms. Current understanding cannot quantitatively design waste forms or predict durability with confidence, but it is used to advantage qualitatively in responding to problems that do arise.

All of the authors were asked to summarize or review their organization's contribution over the years, with liberal use of their own references. As a consequence, most of the papers appear to be reviews and "horn tooting" to a certain extent. In other words, these authors generally acknowledge the work of others in this field, but they were specifically asked to introduce the reader to their organization's achievements and past publications, not to review the subject in general. This also led unavoidably to some overlap among the papers.

In Chapter 1, Prof. Glasser introduces the chemistry of cement pastes and some of the effects of waste constituents that he has observed on this chemistry. In Chapter 2, Jesse Conner summarizes the well-known effects of pH on hazardous waste constituents and the effectiveness of solidification/stabilization based on his long experience. This chapter is a summary of his book, to a certain extent. Chapter 3 reviews some of the extensive work done at the Materials Research Laboratory at Pennsylvania State University, focusing on the chemistry but from the perspective of materials science.

Chapters 4 to 8 do not ignore chemistry, but focus mainly on microstructure and microstructural techniques. Chapter 4 gives an introduction into microstructural techniques, giving a brief description and the potential of these techniques. Chapter 5 discusses the application of these techniques by a group in Alberta, Canada, including the Alberta Environmental Centre, the University of Alberta, CANMET, and the Alberta Research Council. The development of a technique for studying cementitious waste forms by transmission electron microscopy and scanning transmission electron microscopy is particularly interesting. Chapter 6 reviews some of the accomplishments in this field by Louisiana State University (LSU). Chapter 7 reviews the contributions by the University of New Hampshire. Chapter 8 summarizes the contributions from Lamar University.

Chapters 8 and 9 do try to bridge the gap between chemistry/microstructure and leaching. Chapter 8 focuses mainly on the chemistry and microstructure but also addresses leaching. Chapter 9 reviews the attempts at Texas A&M University to include chemical equilibrium and solubility with the diffusion mechanism to model the leaching behavior of cementitious waste forms.

This book is intended to be a statement of what the current state of the art is on the chemistry and microstructure of solidified waste forms. The book can demonstrate to those unfamiliar with cementitious waste forms how much is truly understood about this topic. The book also informs knowledgeable workers in the field about the experiences and accomplishments of fellow workers and provides a wealth of references, as well as some observations presented for scrutiny, which are subject to interpretation.

Roger D. Spence received his B.S. in Chemical Engineering from Virginia Polytechnic Institute and State University in 1971 and his Ph.D. in Chemical Engineering with a minor in Environmental from North Carolina State University in 1975. His Ph.D. dissertation topic was the development of polymeric barriers as sulfur dioxide stack monitors.

Dr. Spence is a member of the Tau Beta Pi and Phi Kappa Phi honorary societies and is a member of the ANS-16.1 committee that is responsible for updating and recertifying the ANSI/ANS-16.1 leaching procedure for solidified waste forms. Dr. Spence participated in workshops to develop a hazardous-waste leach test for Canada, to issue a guidance document on solidification/stabilization for the U.S. Environmental Protection Agency, and to advise the U.S. Department of Energy on research needs for environmental restoration. In addition, he was part of a panel that reviewed solidification/stabilization and thermal treatment environmental restoration development projects for the Office of Technology Development of the U.S. Department of Energy.

Dr. Spence has worked for Oak Ridge National Laboratory since 1975. During his first 10 years, he conducted research and development on advanced separation techniques, advanced nuclear fuel cycles, nuclear fuel reprocessing, and nuclear reactor safety. Since 1985, Dr. Spence has conducted research and development on solidification/stabilization of low-level radioactive waste and hazardous waste, the leaching and leaching theory of solidified waste forms, and the *in situ* solidification/stabilization of contaminated soils, sediments, and shallow land burials. During the course of his environmental restoration work, he has performed a grout injection into a shallow land burial filled with low-level radioactive waste; developed grout formulations for several tasks, including *in situ* grout injections and the solidification/stabilization of several waste types, including liquids, solids, and sludges, ranging from low-level radioactive waste to hazardous wastes contaminated with metals, organics, or both;

studied the polymer impregnation of cements and concretes; and studied the separation of paint from plastic media blasting waste and the low-temperature ashing of this waste.

During his tenure with Oak Ridge National Laboratory, Dr. Spence has received two IR-100 awards, received one patent on polymer-impregnated concrete, filed another patent application on the low-temperature ashing of plastic media blasting waste, and has received three local awards for technical writing. He has published several articles and reports, and presented several papers on his various research topics.

CONTENTS

1 CHEMISTRY OF CEMENT-SOLIDIFIED WASTE FORMS

F. P. Glasser

1.1. ABSTRACT

The complex chemistry, mineralogy, and microstructure of cements is reviewed. Chemical as well as physical factors contribute to the immobilization potential for toxic materials of cement-based systems. Four chemical factors responsible for this potential are discussed: sorption, precipitation, lattice inclusion, and reaction with cement components to form solubility-limiting phases. The internal redox potential of cement matrices also plays an important role in determining speciations and, hence, specific interactions with waste components.

On account of the large number of system variables and the paucity of the database, it is not always possible to predict and explain the immobilization potential in mechanistic terms, but case studies of uranium and iodine in cement illustrate how focused research can provide a scientific basis for determining immobilization mechanisms. Finally, the environmentally conditioned factors which control the durability of cements and long-term rates are examined.

1.2 INTRODUCTION

General Considerations

It is sometimes argued that waste materials are an avoidable problem; that new and improved technology, coupled with new strategies for reuse and recycling, will eliminate waste. New industries and new technologies will

0-87371-748-1/93/$0.00+.50
© 1993 by Lewis Publishers

undoubtedly reduce waste arisings: the profligate slag heaps so characteristic of heavy industry in previous decades will no longer grow, at least not in developed countries. Recycling technology will also improve, particularly in economically advanced economies, where social factors will encourage paying a "green premium" for waste mitigation and conditioning strategies, some of which may not be justifiable in strictly economic terms.

Nevertheless, intractable wastes will continue to arise. It should not be supposed that all new technologies are pollution free: an example is the generation of metal-rich incinerator ashes from modern refuse-treatment plants. Therefore, strategies still need to be devised to cope with wastes, although the amount and character of wastes may be undergoing changes.

Organic wastes are in some respects easier to cope with than inorganic; their molecular structure can frequently be altered to make them less hazardous or innocuous, as, for example, by total combustion. Only in a few cases can inorganics be altered so as to reduced their toxicity; cyanide, which can be oxidized to the less-toxic oxycyanide is an example. However, the inorganic elements cannot be altered and only limited control can be exerted over their toxicity by changes to their speciation. An acceptable strategy for disposal of nontransmutable inorganics will have to be defined within a framework of statutory requirements, but in the main objectives of acceptable requirements will include control of discharge limits and dispersion to the environment. Success in attaining this goal will usually be monitored by analysis of soil, of drinking water, of the atmosphere, etc.; successful conditioning treatments will enable defined release limits to be achieved.

Application of Cements to Solidification Processes

A rapid appreciation of the advantages of cement-conditioned formulations can be obtained by considering the advantages of solidification and the type of matrices available. If mobile dusts, liquids, and sludges are converted to solids, the ease with which they can by physically dispersed to the environment is greatly reduced. Cements thus have a role in solidification; moreover, they have a useful barrier potential: engineers and the general public accept the use of cement-based materials for containment purposes, e.g., in dams or grout curtains.

Purely physical confinement of wastes can be achieved by a wide range of solidification matrices, including organic polymers, asphalt, etc., as well as hybrid materials such as polymer-inorganic cement blends, impregnated materials, or bitumen-cement emulsions. However, cements differ from chemically passive matrices, such as polymer, by providing a significant chemical immobilization potential for many inorganic waste species. While this presentation will concentrate on the chemical aspects of the potential, the reader should be reminded that both physical and chemical aspects of cement behavior are important and, moreover, that physical and chemical mechanisms of immobilization will normally interact: hence, the microstructure, porosity, permeability, and mineralogy of cement materials are also of concern.

1.3. NATURE OF CEMENTS

General

In the ancient world, lime was extensively used as a binder. $CaCO_3$, limestone, is abundant and when "burned" to ~850 to 900°C, it decomposes to CaO, lime, with evolution of gaseous CO_2. The resulting product is powdery or only loosely coherent but reacts readily with water, liberating much heat, to form "slaked lime", $Ca(OH)_2$. This material, mixed with sand, forms a weak mortar which has a high proportion of open porosity. $Ca(OH)_2$ is also somewhat soluble: approximately 1.1 g l^{-1} at 18°C, giving a strongly alkaline saturated solution with a pH of 12.4. On account of their solubility and open porosity, lime mortars are not very durable in environments where they are subject to leaching.

Thermally activated limestones containing a proportion of clay have long been known to make cements more durable than pure limestone. During firing, the clay minerals lose water to form ill-crystallized, fine-grained anhydrous aluminosilicates. When the resulting material is ground, mixed with water and allowed to set, its aluminosilicate content reacts with some of the $Ca(OH)_2$ to form a silicate gel, usually designated C-S-H (shorthand: C = Ca, S = Si, H = H_2O). Thus, dehydroxylated kaolin, $Al_2Si_2O_7$, reacts with slaked lime:

$$Ca(OH)_2 + Al_2Si_2O_7 + H_2O \rightarrow C - S - H$$

The equation is incompletely balanced; the C-S-H product has a variable Ca:Si ratio and, moreover, Al reacts to form other phases which are not shown. However, C-S-H is a central product of reaction which controls the bonding in Portland and modified Portland cements and is responsible for their superior durability. The C-S-H thus formed tends to fill pores, enhances bonding, and imparts more strength to the mix than if it consisted only of $Ca(OH)_2$. Moreover, it is more resistant to leaching than $Ca(OH)_2$ because it dissolves incongruently, leaving a passive and semiprotective residue of rather insoluble $SiO_2 \cdot nH_2O$ (hydrated amorphous silica); thus, while the C-S-H in cement has a Ca:Si ratio ~1.7, the aqueous phase which contacts it has a steady-state Ca:Si ratio ~40.[1] The water permeability and durability of these so-called natural cements are thus much superior to lime mortars. Natural cements remained in widespread use in the U.S. until the early parts of the 20th century, when they were gradually displaced by Portland cement.

Portland Cements

The composition of modern Portland cement evolved from that of natural cements. At the higher burning temperatures used in Portland cement production, approximately 1450°C, all the SiO_2-containing components (clays, quartz, etc.) react to achieve, very nearly, an equilibrium phase assemblage. Modern Portland cements are proportioned to consist of two anhydrous calcium silicates, Ca_3SiO_5 and Ca_2SiO_4, while Al_2O_3 and Fe_2O_3 — the other components — principally are combined as calcium aluminate and calcium aluminoferrite. The batching and calcination should ensure that silica is chemically combined as Ca_3SiO_5 and Ca_2SiO_4, while free CaO is reduced to near zero. The resulting product, which partially melts and sinters during calcination, is known as clinker. This well-indurated clinker is mixed with 2 to 5% gypsum, $CaSO_4 \cdot 2H_2O$, and ground to a fine powder having a specific surface of 3000 to 4000 cm^2 g^{-1}: the product is marketed as "Portland cement". The purpose of adding gypsum is as follows: ground clinker would quickly stiffen when mixed with water, but adding sparingly soluble sulfate delays initial set for 1 to 2 h. This ensures that fresh, wet mixes have a period of plasticity, during which they can be handled or worked.

Commercial Portland cement is marketed primarily for conventional civil engineering and building applications for mixing with sand and gravel, whereas, in waste conditioning processes, it may be used as part of a very different formulation. Therefore, knowledge of cement properties and reactions can be transferred, but only with caution, from civil engineering to waste conditioning.

Hydration Reactions

The set reactions of cement are complex and are still the subject of debate. However, Figure 1-1, which shows the overall course of hydration kinetics, is generally accepted. After an initial burst of heat evolution, during which little hydration occurs, an induction period is encountered. This is necessary to retain plasticity. Thereafter, set occurs as hydration reactions accelerate until, by approximately 28 d at 5 to 30°C, roughly two thirds of the cement will have hydrated. Figure 1-1 does not differentiate between the various solid products of hydration but it suffices to show the apparently smooth curve marking the overall progress of hydration. In the period beyond 28 d, hydration will still continue provided moisture is conserved, albeit at a diminishing rate, until by 1 year 95 to 98% of the cement will have hydrated. The two main products of hydration are $Ca(OH)_2$ and C-S-H. Most modern cements yield about 20 to 25% $Ca(OH)_2$ and 60 to 70% C-S-H; other solid phases comprise perhaps 5 to 15% of the total solids. Table 1-1 lists the crystalline solids. However, it is noteworthy that the principal hydration product, C-S-H, is nearly amorphous to X-ray diffraction. The solid-phase assemblage, especially the $Ca(OH)_2$

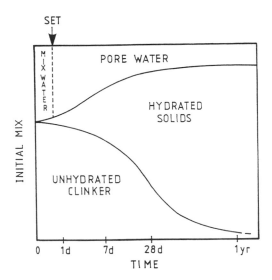

Figure 1-1.   Schematic, showing the overall course of cement hydration. As the mixture hardens, mix water is gradually transformed to pore water: this is indicated by the vertical dashed line.

Table 1-1. Crystalline Components of Set, Hydrated Cements

Designation or Abbreviation	Approximate Oxide Formula	Notes
AF_t	$Ca_6Al_2O_6(SO_4)_3 \cdot 32H_2O$	May contain Si substituted for Al, CO_3^{2-} for SO_4^{2-}, etc.
$\begin{cases} AF_m \\ C_4AH_{13} \end{cases}$	$\left. \begin{array}{l} Ca_4Al_2O_6(SO_4) \cdot 12H_2O \\ Ca_4Al_2O_7 \cdot 13H_2O \end{array} \right\}$	Form solid solution
Hydrogrossularite, or hydrogarnet	$Ca_3Al_{1.2}Fe_{0.8}SiO_{12}H_8$[a]	Typical composition: iron not essential
Stratlingite, or gehlenite hydrate	$Ca_2Al_2SiO_9 \cdot 8H_2O$	
Portlandite	$Ca(OH)_2$	
Hydrotalcite	$Mg_4Al_2(OH)_{14} \cdot 2H_2O$	Mg/Al ratio variable

[a] Apparently a solid solution based on $Ca_3Al_2O_6 \cdot 6H_2O$. The composition shown is from an aged Portland cement: (1) Copland, L. E., D. L. Kantro, and G. Verbeck. 1960. 4th Int. Cong. on the Chemistry of Cement (Washington, DC) Vol. 1, 429; (2) Taylor, H. F. W. and D. E. Newbury. 1984. *Cement Concr. Res.* 14: 565.

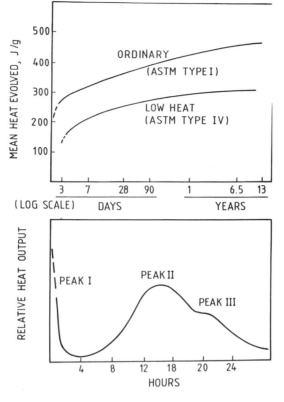

Figure 1-2. Heat evolution of Portland cement showing (top) cumulative curves for two cements and (bottom) output for an ASTM type I cement over the first ~24 h.

content and the Ca:Si ratio of the C-S-H, can be significantly altered by the presence of supplementary cementitious materials, the nature of which will be discussed subsequently. In normal cements, however, the C-S-H has a Ca:Si ratio ~1.7 and its solubility properties are chemically identical with those described in conjunction with natural cements. Of course, the solid matrix also entraps an aqueous phase since, in practice, more water is added to ensure plasticity than is necessary to achieve complete chemical hydration of the solids.

Some of the complexities of hydration can be appreciated by examining Figure 1-2, showing overall heat evolution and presenting the rate of heat evolution over the first few days of reaction. The rate of heat output is, surprisingly, very uneven as a function of time. An initial burst of heat evolution occurs within seconds of mixing, giving the so-called exothermic peak I which is attributed to the heat of wetting of the solids. A period of reduced thermal activity follows, after which heat is again evolved rapidly, giving at least two maxima within 72 h, the so-called peaks II and III. Peaks

II and III are attributed to rapid formation of the calcium aluminosulfate hydrate, ettringite, and the hydration of calcium silicate clinker minerals, Ca_3SiO_5 and Ca_2SiO_4, to form C-S-H and $Ca(OH)_2$, respectively, Thus, the mineralogical changes which attend hydration occur at unequal rates and are, in part, sequential.

The overall setting and strength gain reactions are strongly exothermic; if heat is prevented from escaping, or if large masses of cement are set under semiadiabatic conditions, heat evolution may result in unacceptable temperature rises. For example, a 200-liter drum of neat hydrating cement may experience centerline temperature excursions of more than 80°C above ambient and steam may be evolved. If excessive thermal cracking of set product is to be avoided, temperature excursions of this magnitude should be avoided and limited to less than 30 or 40°C. Waste conditioning treatments must take into account the physics of heat generation and transfer, as well as the chemistry of the hydrating system. Heat evolution occurs sufficiently slowly after the first 48 to 72 h, so that further heat evolution does not cause problems except in very large pours, as in the 5×10^6-liter vaults intended for radioactive wastes at Hanford, WA.

The water content, usually defined in terms of the water:cement (w/c) weight ratio, is critical to the physical properties of the product. Since wastes are often liquid or wet solids — and for economic reasons it is desirable to achieve as high a waste loading as practicable — it appears superficially attractive to use high w/c ratios. However, as will be shown, the use of high ratios can lead to a decline in the physical properties. The general relationships between w/c ratio and properties are well established. For simplicity we consider a closed system, i.e., one in which no water is lost or gained. Briefly, for a neat cement, w/c weight ratios <0.25 will contain insufficient water to achieve complete hydration, nor will the resulting mix flow unless special chemicals, known as plasticizers, are added. The critical ratio for normal plasticity or flow is about 0.30, but, at this w/c ratio or above, a chemical excess of water will be present after cement hydration is complete. The chemical excess of water over that required for hydration will remain trapped interstitially within the set mass, although without unduly increasing the permeability of the hardened product. Higher w/c ratios, above 0.35, will give increased fluidity but will lead to the inclusion of progressively more trapped pore water and, as a consequence, have greater permeabilities and decreased final strengths. A detailed treatment of porosity-permeability relationships lies beyond the scope of this paper. However, a recent review of these data is available[2] and mathematical treatments of pore connectivity have been made.[3]

At very high w/c ratios, segregation of mix constituents may occur before set can be achieved. Since Portland cements are "hydraulic", i.e., they set under water, a cement product will continue to set under segregated water, known as bleed water. The presence of bleed water is, however, an embarrassment in many immobilization processes since it is not only mobile, but may well also contain a significant content of soluble or particulate toxics.

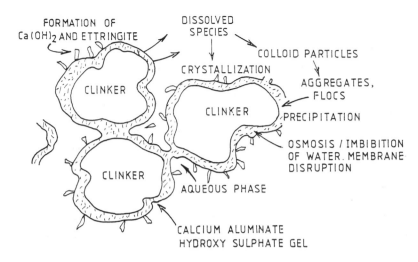

Figure 1-3.   Schematic view of hydration processes occurring during the first few hours or days.

Hence, the practical window of w/c ratios for operating cement-based conditioning processes is frequently rather small. The presence of soluble waste constituents, of colloids, etc. may affect the rheology of the wet mix and hence the numerical limits of the "window" in which conditioning processes can operate, but the general principles described above will apply to all cement-based systems.

The general nature of Portland cement hydration appears to involve both dissolution-precipitation (known as through-solution) reactions as well as direct attack of water on grains of anhydrous cement. Through-solution mechanisms, in which anhydrous solid cement components dissolve, supersaturate the aqueous phase, and finally reprecipitate, mainly as hydrates, are probably of greatest importance during the initial set and hardening: solution-precipitation is very effective in transporting material and gelling the aqueous phase. Figure 1-3 summarizes some of the processes involved in early hydration. In making this interpretation, the writer is aware that some divergence of views exists concerning the nature and relative importance of the several processes shown, so the diagram is intended as a summary guide to the literature: hydration has been authoritatively reviewed in a recent book.[4] Pictorially, the diagram shows the state of a cement system a few hours after mixing, by which time much of the gypsum or soluble sulfate will have dissolved and reacted. By this stage, the clinker grains have become coated with a layer of nearly amorphous gel-like material. The coating is added to by dissolution of clinker and subsequent precipitation of fresh hydrates. In part, reaction occurs by direct attack of water on the grains and in part indirectly, through solution, by colloid formation, ripening and coagulation with subsequent agglomeration, and deposition of newly formed solid. The dissolution processes maintain a supersaturated solution, most notably with respect to calcium, during this phase of hydration. Meantime, the initial hydrate coatings established around clinker grains act to some extent as membranes through which diffusion must occur. The stability of the early-formed coatings is often disrupted, mainly by partial crystallization of sulfate-rich gel to yield crystalline ettringite, as well as by decreasing solution supersaturation. By 6 to 12 h, the gel layer continues on average to thicken, but it also densifies and gradually transforms in composition; with increasing cure duration it contains less Al and SO_4^{2-}, and more silicate.

One of the principal products of later hydration, beyond 6 to 24 h, which develops strength and has good space-filling properties, is a calcium silicate gel. The gel phase in Portland cement has a maximum Ca:Si ratio of ~1.7, whereas the Ca:Si ratio of the cement in ~2.5; the chemical excess of Ca which cannot be accommodated in C-S-H appears mainly as $Ca(OH)_2$.

Like most gels, cement gelation processes are sensitive to the ionic strength and chemical constitution of the aqueous phase. The presence of soluble salts alters the rate of set and the morphology of hydrated products while the presence of species which complex calcium strongly, or which precipitate it by forming noncementitious phases, may delay set or have the even more drastic effect of preventing final set and strength from ever being achieved. Substances inimical to set include those which depress Ca solubility, e.g., borates and phosphates, as well as those which form soluble complexes with Ca, e.g., citrate, EDTA, and simple saccharides; the list of interferences is potentially very long.

There are, therefore, limits to the tolerance of cement systems set by permissible water contents and temperature excursions during solidification, as well as by the tolerance of the setting system for foreign materials. These constraints further limit the windows within which cement conditioning processes can operate. Conventional cement technology often enables us to suggest acceptable limits for water contents and temperature excursions, although rather less data are available for predicting the impact of soluble materials on set and strength gain, as well as on the nature of the ensuing reactions between cement and embedded solids. Consequently, it is frequently necessary to establish empirical tolerance values for specific waste streams in cement matrices or to evaluate modifications to the cement formulation, so as to overcome potential incompatibility problems.

1.4. PHYSICAL PROPERTIES

Strength

Because of the extensive structural uses of cement concrete, a vast literature exists on the strength of concrete. However, most of this literature, as well as a large body of national specification and codes, is not necessarily relevant to waste conditioning.

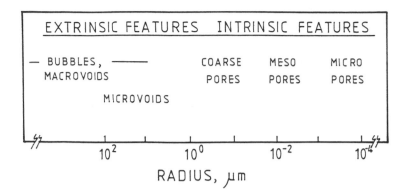

Figure 1-4.   Classification scheme for the porosity of cement paste.

Conditioned wastes may require shipment and handling, in which case appropriate strength requirements may have to be imposed, but these may well be less severe than for good structural concrete. In many applications such as landfills, grout curtains, etc., very low strengths, on the order of a few megapascals, may well be acceptable. It is suggested therefore that any structural elements, e.g., load-bearing concretes, will have to conform to appropriate national codes, but that a wide range of other materials, mainly of lower strengths, may be acceptable for nonstructural applications such as waste solidification treatments.

Pore Structure of Cements

The pore sizes of cement cover a very wide range of sizes. Figure 1-4 shows the spectrum of sizes and gives an indication of the origin of porosity in certain ranges in a well-matured cement paste. This pore distribution is not achieved in fresh cement, which will inevitably contain a higher free water content, and hence more pore space, but will be obtained after prolonged moist cure. Persistent larger pores, >1 μm, tend to be artifacts of formulation and processing; for example, air bubbles may be entrained during mixing. Indeed, some dispersible organics greatly assist air entrainment by stabilizing small air or gas bubbles. In wet mixes containing particulate matter, sedimentation of particulate substances may occur, leaving behind water-rich annuli which may never fill completely with solid cement substances; these are, effectively, pores. The larger pores can be considered as extrinsic features, inasmuch as the amount and distribution of porosity in this range depends on formulation, preparation, and cure. Thus, mature pastes, made to low w/c ratios and well cured, rarely contain significant porosity above 1 μm unless they are air entrained. Below this, however, lies the range of intrinsic porosity. This is rather arbitrarily divided into coarse, meso-, and microporosity. The coarser paste pore microstructure tends to reflect the original packing of cement grains against each other and against other surfaces, e.g., those furnished by container walls and waste particulate material, as well as the initial w/c ratio. The cement grains floculate during the early stages of set, but the semirandom geometry of flocs is far from being close packed, with the result that larger voids never become completely filled with cement hydrate product. Micropores, on the other hand, are an intrinsic feature of the C-S-H gel. The gel itself may be viewed as a coherent assemblage of colloidal particles, including lamellae and foils, the packing of which gives rise to micropores. The mesopores arise from a combination of all the above intrinsic features, as well as from the subdivision of larger pores by growth of hydration products in cavities; growth products commonly include ettringite and $Ca(OH)_2$.

A wide range of experimental methods has been used to investigate the total porosity and pore size distribution. In general, progressively more sophisticated methods are required to determine progressively finer pores. Thus, simple impregnation techniques may suffice to determine coarser pores while finer pores have to be determined by mercury intrusion at controlled increments of applied pressure, by N_2 sorption, or by neutron scattering. The latter technique does not require previous drying of the sample but does of course require a neutron source! Each technique has its advocates, but fortunately good general agreement is obtained about the large pores and mesopores, which together dominate the observed diffusional properties. In general, mercury intrusion remains the primary tool for characterizing the pore structure. The method has been reviewed recently, together with algorithms which stimulate the intrusion process in two dimensions.[5]

It is appropriate to add several cautions about mercury intrusion. It will not measure the very finest, or gel porosity, nor will it measure closed porosity. Fortunately, neither of these factors appear to be too important in the present context; as pore diameters fall below 1 nm, it normally becomes adequate to treat the micropore network and C-S-H together, as though they constitute a homogeneous material. Moreover, hidden porosity contributes little to observed diffusion profiles.

Rapid diffusion through cementitious matrices is associated with larger pores and their openness, or connectivity. For this reason, measurement of the large and mesopores, together with the extrinsic open porosity, provides a useful measure of cement quality. Figure 1-5 depicts some typical intrusion scans for a well-made paste. As moist curing proceeds, the total porosity decreases and the median pore size shifts to progressively smaller regimes. This is quite characteristic of a maturing paste; continuing hydration of clinker progressively fills and subdivides pores. These observations help explain why it is important to establish the intrinsic leach properties of a cement after a steady state has been reached. If intrinsic leaching is measured while curing is still progressing actively, the leach data reflect a mixture of two factors: continued pore structure reduction and refinement, and intrinsic retentivity. Of

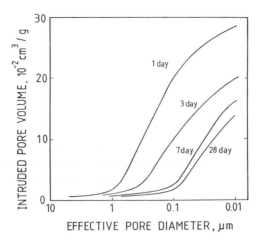

Figure 1-5.   Mercury intrusion porosimetry curves for a typical good quality cement paste wet cured for various periods at ~20°C.

course, there may be good reasons why, in certain applications, leaching must be measured after very short cure durations.

However, in general, if the steady-state properties of the material are to be measured, a moist cure of, say, 60 to 90 d is recommended prior to commencing leach testing. One of the least-known factors, but one which is important in present context, is the relationships among porosity, permeability, and microstructure. Quantitative descriptions of cement microstructures are as yet in their infancy but qualitatively the microstructure is well explored. Figure 1-6 illustrates a "normal" microstructure, in this instance of a cement paste made to a w/c ratio of 0.35 and moist cured for 180 d. The specimen has been prepared for microscopy by fracture, which emphasizes the perfect cleavage of blocky $Ca(OH)_2$ crystals. Hence the multiply cleaved $Ca(OH)_2$ platelets are perhaps the most conspicuous feature. In this paste the $Ca(OH)_2$ content, determined by thermogravimetry, was approximately 24%. However, the bonding matrix consists primarily of a relatively featureless C-S-H gel. The gel still tends to cluster, each cluster marking the approximate site of a former clinker grain. The outer portion of the gel, known as "outer hydrate" product, has a relatively open, low-density structure, while inner product, at the core of grains, is denser and more massive. Many investigators have described pseudocrystalline morphologies such as fibrillar structures and crumpled foils, especially in outer product, but these features are less common in low w/c ratio pastes and are generally absent from the microstructure of inner hydrate. A few fine rod or needle-like crystals of ettringite are visible; the actual ettringite content of this paste is approximately 5%, by semiquantitative X-ray. The larger pores are clearly visible; they tend to cluster between hydrated relict clinker grains and at regions of poor fit between C-S-H gel and $Ca(OH)_2$ plates. In general, while optical microscopy is satisfactory for measuring the amount

Figure 1-6.   Scanning electron micrograph of a typical well-matured cement paste.

of larger voids, electron microscopy, as in the present example, is useful at least to visualize the microstructural role of coarse pores, and perhaps of the mesopores. Quantification of the pore structure by electron microscopy is developing in parallel with the increasing sophistication of image analysis techniques.

The normal Portland cement paste microstructure is capable of considerable modification. Fly ash and slag are frequently used to modify its chemical composition, as well as its microstructure. These supplementary cementing materials are characterized by having a mean particle size finer than that of clinker, which is typically 5 to 25 μm; ground slag is somewhat finer, but fly ash — depending on the source — may have an abundance peak in the 1- to 5-μm range, corresponding to a large increase in geometric surface area relative to cement. Although slag and fly ash may react with water more slowly than cement, the improvement in initial solid particle packing brought about by the introduction of fine particles, and the corresponding bimodal particle size distribution, eventually leads to a denser matrix being achieved in well-cured pastes, other factors being equal. Moreover, many fly ashes are rich in Al_2O_3 and SiO_2, and these components, especially SiO_2, react with and rapidly reduce the amount of residual $Ca(OH)_2$ in cured compacts. If present in sufficient quantity, siliceous fly ash, and to a lesser extent slag, may also lower the Ca:Si ratio of the C-S-H gel, from its normal value of approximately 1.7 in unmodified cement possibly to as low as 1.0 to 1.2. Changes in Ca:Si ratio, as will be shown, affect the sorption properties of the gel.

1.5. IMMOBILIZATION POTENTIAL

General Overview

Portland cement has a number of chemical features that are responsible for its widespread use in toxic waste immobilization processes. It is tolerant of wet material — indeed, water is required for set to occur — it is not flammable and is durable in the natural environment. Moreover, it can be used as an activator for other potentially cementitious materials, e.g., glassy slag or fly ash. These supplementary cementitious materials eventually become an integral part of the cementitious matrix with the result that they provide a matrix which utilizes one waste (fly ash, or slag, or mixtures of the two) to immobilize others. Thus, there are distinctive chemical features which make cement-based systems attractive.

Some of these aspects are summarized in Table 1-2; the underlying mechanisms which give rise to these features will be explored subsequently. It is important to note that the present extent of our knowledge in this area varies; conventional cement technology is not normally concerned with toxic elements and little incentive has existed until recently to determine specific interactions

Table 1-2. Chemical Immobilization Potential of Cements

- Absorption of ions into, and adsorption on, high surface area C-S-H
- Precipitation of insoluble hydroxides, owing to high alkalinity
- Lattice incorporation into crystalline components of set cements
- Development of hydrous silicates, basic calcium-containing salts, etc. which become solubility-limiting phases

between cement and toxics, including simple ions as well as more complex species. Numerous processes — often patented — employ cement either singly or in conjunction with other alkaline media, e.g., calcium hydroxide and/or sodium silicate, but these frequently lack any verification of the specific mechanisms whereby their claimed performance is achieved.

Performance data are often based on leach testing. The results obtained often point to low, slow release rates being achieved. This is desirable in the context of their intended application. However, while these results broadly indicate successful application of cements and modified cements, they do not form a quantitative basis on which the specific mechanisms of immobilization can be quantified. Leach test results can ultimately be explained only on the basis of diffusion theory combined with a knowledge of the intrinsic chemical features of cement matrices which promote — or inhibit — immobilization. These chemical features and cement-waste component interactions will be explored.

Internal pH

The persistence of an aqueous phase, as pore fluid in set cement, has been noted. This aqueous phase pervades the matrix, with which it is in intimate contact. Therefore, cements may be described by functions which, at first sight, might seem inappropriate for a solid; for example, the application of pH concepts. The alkaline nature of pore solutions has been noted. Not only is $Ca(OH)_2$ somewhat soluble, but so too is C-S-H, the gel substance of cements. Phase diagrams showing the coexistence regions in the $CaO-SiO_2-H_2O$ have been determined at ~20°C[6] and free energy functions tabulated for the solid phases.[7] For purposes of pH control, Figure 1-7 may be used to show how Ca- and Si-based compositions can be divided into several regions in which a liquid phase coexists with particular solid(s). In this model, an aqueous phase coexisting with two solid phases is seen to have a fixed composition at a specified temperature and pressure. The behavior of real, more chemically complex cements broadly reflects this phase rule restriction. When cement or mixtures of cement and hydrated lime are used, the bulk composition of the solids lie in region I, with the result that the aqueous phase composition is constrained to lie on the solubility curve, at its intercept with an extension of the lime a − r.

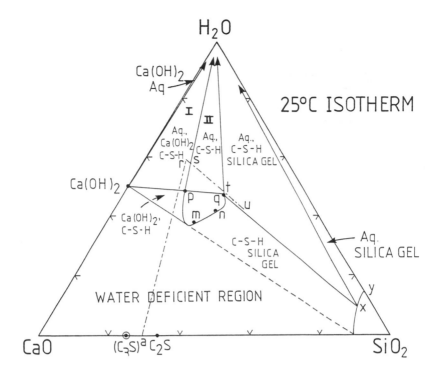

Figure 1-7.   Schematic of the CaO-SiO$_2$-H$_2$O system showing hydration products obtained at ~25°C. The region of C-S-H compositions lies in the area p-q-n-m. If a cement, a, is mixed with water the bulk composition is shifted to r. The progressive addition of a blending agent, assumed for convenience to be pure SiO$_2$, would shift the composition along the dashed line s-t-u. This affects the pH: see Figure 1-8.

In practice, the composition of the aqueous phase at this point is such that the solubility of Ca, approximately 40 mM, greatly exceeds that of SiO$_2$, approximately 1 mM, with the result that the composition and pH of the aqueous phase do not differ much from that which would be conditioned by the solubility of Ca(OH)$_2$ alone; the solution at point a has a pH close to 12.4 at 18 to 25°C. Note that the coexistence of Ca(OH)$_2$ and high Ca:Si ratio C-S-H not only produces a high pH but also buffers the pH. Thus, the internal pH of the aqueous phase does not depend on the *amounts* of Ca(OH)$_2$ and C-S-H which are present; rather it depends on the *continued presence of both solids*, and on the maintenance of a *steady state, quasi-equilibrium between solid and aqueous phases*. This is an important and often-overlooked consideration: while infinite variations can occur in the mix ratios of cement, hydrated lime, and water and, while cement compositions may vary, many practical situations can be expected to give rise to closed system behavior in which broad ranges of solid compositions will coexist with an aqueous phase having an essentially fixed composition; hence, the importance of the solution invariant composition, especially of point a, to waste conditioning.

Three factors could potentially affect this analysis. One is the presence of additional components known to be present in cement, including, for example, alkalis, alumina, and iron oxides. The second is the content of supplementary cementitious materials, reaction of which removes $Ca(OH)_2$ and changes the bulk Ca:Si ratio of the cementitious matrix. Third, in the longer term, environmentally conditioned reactions leading to degradation of cement performances have to be considered; the system becomes open with respect to mass transport. Leaving aside the latter for the moment — open systems are of great importance, but difficult to treat without first having established the behavior of a closed system — these factors can be dealt with in turn. With respect to the other minor components, they can be divided into two classes. The first class includes cations such as Al, Fe, Mg, Mn, etc., which are characterized by forming compounds which, by and large, are found experimentally to have little effect on Ca solubility in conjunction with the principal buffering pair, $Ca(OH)_2$ and C-S-H. Therefore, the occurrence of this class of ions, distributed between various phases, does not significantly affect the pH of the system; the simple model, developed for the calcium component is still applicable. The second group of minor components comprises mainly soluble ions, e.g., sodium and potassium. These do not form compounds at normally encountered concentrations, although sodium replaces Ca in C-S-H to a limited extent and is thus partitioned between aqueous and solid phases; partitioning data are known quantitatively for most commonly encountered compositions.[8] It can generally be concluded that sodium, and especially potassium which is rather less soluble in C-S-H than sodium, tends to raise the pH above that of the buffer system. This pH elevation is obtained even in the presence of other anions, e.g., SO_4^{2-} or Cl^-, at least at low concentrations, because these anions react with cement components and become occluded within solids, thereby leaving OH^- as the principal counter ion in solution, This point deserves amplification because predicting the impact of salts on cement pH is a common source of confusion in the literature. For example, it is frequently argued that, since salts such as NaCl or Na_2SO_4 are neutral, their presence in cement systems must be to either lower its pH or else to leave its pH unchanged. This is not so, because of the ability of cement systems to remove a broad range of commonly encountered anions from solution, replacing them by OH^-. Hence, soluble alkali in cement, irrespective of source, tends to act like NaOH or KOH and thereby raise the pH. This, in turn, depresses calcium solubility because of the common ion effect. Clearly, this mechanism has chemical limits but these limits are, as yet, not well established and would appear to vary markedly for each specific anion. However, the general tendency of alkalis, irrespective of the source, to raise the system pH is undoubtedly true.

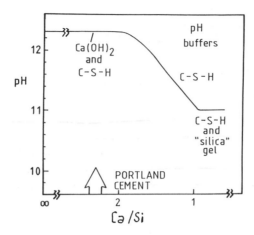

Figure 1-8.   Schematic, showing how the pH of the aqueous phase relates to the mean Ca:Si ratio of cement.

Since alkalis are not well bound into cement solids, the processes of diffusion and leaching which occur in open systems will lead to loss of alkali and an initial decline in pH. However, any immediate decrease arising from alkali leaching will be slight and will be arrested at the level at which the C-S-H/Ca(OH)$_2$ buffer becomes operative. This buffer, on account of the fairly low solubility of the relevant solids and the high total mass per unit volume of Ca(OH)$_2$ and C-S-H, is likely to remain operative over long time periods. Hence, a conservative assumption in predicting future cement performance is to take the initial pH as being about 12.4 at 18°C.

The presence of supplementary cementitious materials (slag, fly ash, etc.) leads to their reaction with the Ca(OH)$_2$ of cement. Fly ash, which is often high in SiO$_2$, has the greatest potential for reaction with Ca(OH)$_2$. If the proportioning of the cementitious matrix is such that portlandite is totally consumed, only high ratio C-S-H will remain to control pH. Slag and ASTM Class C fly ash, with their lower SiO$_2$ contents, are less effective on a weight basis in reacting with Ca(OH)$_2$, but do generally consume Ca(OH)$_2$. Once all the Ca(OH)$_2$ has been consumed, further reaction will lead to a decline in the Ca:Si ratio of C-S-H, with the result that the pH will also decline as compositions enter region II (Figure 1-7). The supplementary construction, Figure 1-8, shows how the pH will decrease in this region; the decline is relatively slow since all C-S-H's are alkaline. Thus, as long as C-S-H is present to replenish the aqueous phase by incongruent dissolution, the pH will remain high, above ~11.

Some wastes may also exhibit pozzolanic properties. Perhaps the most notable group of wastes — and one for which no significant body of data exists — are glassy incinerator ashes. If these materials do not have a high specific surface, their pozzolanic properties may not be so evident in the short term as with high surface area materials such as fly ash; on the other hand, those ashes which have high intrinsic surface areas may be rapidly pozzolanic.

Internal Redox Potential (E_h)

Cements are normally produced under oxidizing conditions and, as a consequence, most of their iron content is present as Fe^{3+}. Measurements of the E_h of cement pore fluids typically give values in the range +100 to +200 mV. The principal electroactive species in pore fluid are believed to be oxygen dissolved in mix water and perhaps traces of sulfite, SO_3^{2-}, formed by condensation of kiln vapors which contact the cement during cement clinker cooling. The cement solids may additionally contain small amounts of Fe^{2+} and Mn^{2+}, but these remain insoluble at high pH and are not rapidly electroactive. However, manganese and chemically reduced iron are only minor constituents and cements generally lack any active redox couples among their main constituents, i.e., their E_h is not well poised. Hence, other electroactive materials introduced into the system are liable to influence its internal redox state.

The most important supplementary cement materials, fly ash and slag, have different impacts on the E_h. Although fly ash contains significant amounts of unburned carbon and chemically reduced iron, mainly as Fe_3O_4 (magnetite), its chemically reduced phases have little impact on the E_h. Apparently, the unburned carbon and magnetite contents of fly ash are kinetically too inactive to influence the E_h. Blast furnace slags are rather more electroactive: they contain much (1 to 4%) sulfide, S^{2-}, which replaces O^{2-} in the glass network structure. Thus, as hydration proceeds, chemically reduced sulfur is liberated in labile form with the result that the poising capacity of the system is increased markedly relative to OPC; moreover, the E_h decreases to reflect that of the slag chemistry. Under alkaline conditions, some of the sulfur content of slag is soluble in pore fluids where it reacts partly to form thiosulfate, $S_2O_3^{2-}$. The combined solubility of the reduced S species also increases that of SO_4^{2-}, so enriching the pore fluid in total S. However, before the E_h can decrease from its initial, slightly oxidizing value, the limited poising capacity of the OPC must first be overcome. Typically, this requires at least 40 to 50% slag replacement and 30 to 90 d of cure for sufficient reaction to occur between mix components to achieve the low E_h state. Figure 1-9 shows how the redox potential declines as a function of slag replacement. There are some differences in the literature concerning the numerical value of the lower E_h value which will be achieved by slag addition, but it is now generally agreed that correct values can only be determined using pore fluid which has been collected in the absence of air and that low values, in the range −350 to −450 mV, are a correct reflection of the conditioning action of S-containing slag.[9,10]

In practice, real wastes may also strongly condition the redox potential. The presence of organics and of reactive metals, including steel, can be expected

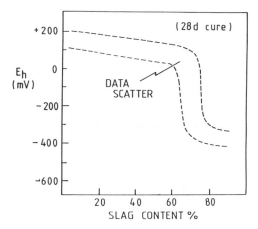

Figure 1-9.   **Schematic E_h-composition diagram showing the impact of slag additions on the pore fluid E_h.**

to lower the redox potential. Thus, cement conditioned by steel will probably achieve an E_h as low as -600 mV, i.e., the E_h which is controlled by the discharge of H_2 at the relevant pH.[11] Organics present in conditioned waste will also presumably lead to a lowered E_h, although little practical work has been done on quantifying their role. However, unless E_h conditioning is achieved by some method of homogeneous addition, such as by using slag replacement in the matrix formulation, it is unlikely that a homogeneous, low E_h matrix can be guaranteed; reaction kinetics between matrix and waste components and the possibility of ingress of oxygen from external sources are balancing factors which may require quantification before the internal E_h can be assessed.

Where reducing conditions can be maintained, it is known that the ionic solubilities of higher-valent species are often much reduced. For example, studies on radioactive wastes disclose that Tc is present in significant concentrations as the soluble pertechnate, TcO_4^-, species in normal Portland cement pore waters, whereas in high slag-Portland blends its solubility, and hence leachability, decrease by several orders of magnitude.[12] In the lower E_h regime of slag cements, Tc is reduced to the Tc^{4+} state which, like those of quadrivalent Ti, Zr, Hf, higher valent Mn, etc., is very insoluble in alkaline solution. Similar considerations probably apply to other, less exotic ions which can be reduced to lower-valent species in the low E_h-high pH regime of slag cement matrices; Mn and Cr furnish examples which are relevant to toxic waste control.

Manipulation of the redox potential within cement thus affords additional possibilities for controlling the speciation and solubility of toxics. The cement matrix itself is only weakly electroactive, so addition of small amounts of an electroactive material intended to control E_h affords considerable scope for tailoring the chemical properties of the matrix. The most favorable conditions

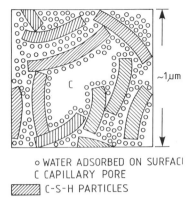

o WATER ADSORBED ON SURFACE
C CAPILLARY PORE
///// C-S-H PARTICLES

Figure 1-10.   Hypothetical reconstruction of the nanostructure of a C-S-H gel.

for immobilization of toxic wastes would appear to lie in achieving a low E_h, so that multivalent metals, typically present as anionic species when in high formal oxidation states, can be chemically reduced in high pH environments to lower-valent, less-soluble cationic species. Of course, the persistence of reducing conditions cannot be taken for granted. However, many saturated disposal environments will also have a characteristically low E_h; clays containing organic materials furnish a common example, as do ground waters previously filtered through a bioactive soil layer, both of which tend to deplete their oxygen fugacity. Thus, it appears realistic to achieve and maintain high pH, low E_h regimes within cement-conditioned materials, even when intended for shallow burial, provided appropriate site conditions exist.

Sorption Potential

C-S-H has a high specific surface area; measured values for N_2 sorption lie in the range of 10 to 50 m^2 g^{-1}. Figure 1-10 shows a view of the nanostructure of C-S-H gel which helps explain the abnormally high surface area. Since we have no structure-sensitive tools with which to depict precisely the atomic arrangement of disordered structures, the representation is of necessity somewhat schematic. Nevertheless, it incorporates many of the known features of the C-S-H "structure", which is depicted as containing blocks of layer-structured calcium silicate material with silicon in a low state of polymerization. This structure gives rise to strong unsatisfied surface charges which cause strong bonding to water molecules, as well as to other C-S-H nanoparticles. The irregular stacking of solid blocks, or domains, each ~10 to 100 nm, creates a large volume of micropores, ranging in equivalent spherical diameter from a few to a few tens of nanometers. The large specific surface thus created, with its high density of irregular hydrogen bonding, creates a strong potential for sorption. Moreover, the surface charge of C-S-H varies with its composition; Ca-rich C-S-H has a positive surface charge and, as a consequence, tends to sorb anions, e.g., sulfate. However, as the Ca:Si ratio of C-S-H decreases, the surface charge gradually lessens, passing through zero at a C:S ratio about 1.2, and eventually becoming negative at lower ratios. Thus, the more SiO_2-rich, low Ca:Si ratio C-S-H's are rather better sorbers for cationic species than are high-ratio material; however, the magnitude of this potential is still small relative to tailored microporous materials such as zeolites.

Cement-based systems require at least a few days of hydration in which to generate significant quantities of C-S-H. Nevertheless, the first-formed C-S-H material does appear to have an especially high chemical reactivity, occluding many otherwise soluble species. It can be speculated that this arises because of the importance of colloid generation and coalescence, with resulting sorption and physical inclusion of toxics, during the early stages of hydration. However, the potential for sorption by through-solution development of C-S-H is limited by the relatively low silica content of cement, as well as its low availability to the aqueous phase during the initial hydration stages. However, the amount of C-S-H formed, particularly during the early stages of hydration, can be powerfully affected by additives which contribute soluble silica. Sodium silicates are frequently used in this context: they are readily soluble, inexpensive, and do not interfere with cement set. The more sodium-rich formulations are alkaline and, upon mixing with acidic wastes, react with and coagulate many metals, precipitating them initially as amorphous hydrated silicates. In conjunction with $Ca(OH)_2$, or with $Ca(OH)_2$ furnished by cement hydration, sodium silicates tend to give a relatively SiO_2-rich C-S-H precipitate which, like hydrous silica, has a strong coagulating effect on many metals, drastically lowering their ability to remain in solution. The nature and stability of these rapidly forming coprecipitates is as yet unknown, but it is probable that, with continuing moist cure, the initially amorphous precipitates will transform to more crystalline phases. Some additional data and speculations on the nature of these conditioning processes will be presented subsequently. Silicate-conditioned materials appear to perform well in laboratory tests of leachability and are the subject of numerous patents.

Cement sorption processes are likely to exhibit broadly the same features as other sorptive systems which have complex chemistries. Competition for anion and cation sorption sites can occur, conditioned by the high pH and consequent high [OH−] and the reserve of other soluble ions in cement. Soluble anions present in cements include mainly OH− but also SO_4^{2-}; soluble cations include mainly Ca^{2+}, Na^+, and K^+. Thus, the performance of "real" C-S-H, where competition for sorption sites can occur, is often poorer than that derived from laboratory simulations in which competing ions — other than Ca^{2+} and OH− — may be excluded. Nevertheless, the "through solution" generation of C-S-H has an important role in removing dissolved species by sorption in or on colloids which subsequently coagulate. This sorptive contribution can be enhanced by promoting through-solution reactions, as by adding sodium silicates.

Crystallochemical Incorporation

Phase Balance

While C-S-H has a relatively amorphous "structure", which defies complete elucidation and quantification by existing structure-sensitive techniques, cements also contain crystalline phases. Table 1-1 lists some of these. Of course, not all of these phases will be present simultaneously. Which will be present depends on the formulation of the cement, including the amount and composition of blending agent or supplementary cementitious material, cure duration, and temperature. We are some way from achieving a complete understanding of the balance of the coexisting phases and its response to cure, even in the presence of "pure" materials, i.e., in cement matrices without added waste constituents. It therefore seems appropriate to treat the individual cement hydrate phases and conclude with brief remarks on their properties and likely occurrence in hydrated Portland cement and its blends: additional data are given in Table 1-1.

Crystal Chemistry

AF_t This calcium aluminosulfate (with some iron substituted for Al) is characterized by having a framework of calcium, aluminum, and SO_4; its framework is extremely open, and water molecules are distributed more or less randomly within structural tunnels parallel to the needle axis of the crystals (crystallographic c). Its unusually high water content and loose bonding implies rapid exchange of H_2O with its surroundings. However, the framework has a low solubility and remarkable chemical stability, despite its low reticular bond density. It presents several possibilities for crystallochemical incorporation. Tetrahedral anions such as CrO_4, AsO_4, etc. may, in part, substitute for sulfate in the framework and the open channels within the framework create additional potential for inclusion of small organic molecules. Reports of channel substitution of polar organics do not, however, appear to be well authenticated.

AF_m This phase tends to form in the presence of limited sulfate availability. Scattered observations show that its structure is apparently more tolerant of spherical, noble gas-type ions that is AF_t. Thus, in radioactive waste streams, AF_m shows better fixation for iodine, in its I^- speciation, than AF_t.

Other Phases $Ca(OH)_2$ appears to have little potential for crystallochemical uptake; the potential for the remaining phases — stratlingite, grossularite,

Table 1-3. Crystallochemical Incorporation of Toxic Waste Materials in Crystalline Cement Phases

Mode of Incorporation	Example
Substitution for calcium	Sr, Ba, Pb
Substitution for hydroxyl	F^-, Cl^-, Br^-, I^-
Substitution for SO_4^{2-} in AF_t and AF_m	IO_3^- etc.
	CrO_4^{2-}, SeO_4^{2-}, etc.
Substitution for Al, Fe	M^{3+}, Cr^{3+}, etc.
Occupancy of channel sites in AF_t	Small organic molecules

gehlenite hydrate, and hydrotalcite — is as yet largely unknown. The role of hydrotalcite perhaps deserves brief comment. Although hydrotalcite is not normally a constituent of Portland cement on account of their low Mg contents, significant Mg may be introduced with blast furnace slag. In NaOH-activated slags, this Mg appears as $Mg(OH)_2$, but with less-alkaline activators — including Portland cement, $CaSO_4$, and $Ca(OH)_2$ — hydrotalcite is favored. Naturally occurring hydrotalcite contains CO_3^{2-}, but slag cements appear to develop the hydroxylated analogue in which $OH^- >> CO_3^{2-}$. Structurally, the hydrotalcite phase consists of a stacking of brucite- and gibsite-like layers and is a representative of the so-called "basic salt" family of structures. Unpublished studies in our laboratory demonstrate that Cr^{3+}-containing hydrotalcites can readily be synthesized and are very insoluble at near neutral and higher pH's. Its layer structure is very flexible, and can incorporate many other di- and trivalent cations, e.g., Ni and Co in place of Mg, Cr^{3+} in place of A, etc.; a wide variety of anions, e.g., C, Br, NO_3, may also substitute for X^- (or X^{2-}). Hydrotalcite appears to be one of the few cases where stability constant data have been reported for selected anion substitutions.[13] Thus, basic salt structures, of which hydrotalcite is a representative, may be an important source of the immobilization potential of cement systems for a wide range of cations and anions.

The overall potential of the cement phases to incorporate ions is summarized in Table 1-3. This table should not be taken as implying that partition coefficients will necessarily be favorable for immobilization in real systems, where competition for exchangeable sites occurs. In most known examples — and relatively few examples are well documented — the partition coefficients between toxic species and competitive cement species are such as to be only moderately favorable for immobilization, especially of anions, given to high [OH^-] regime which exists in cements.

Thus, while selective strategies involving crystallochemical incorporation are possible, the success of such strategies is limited largely for three reasons. One is that the host structures are highly selective, with the result that partition coefficients tend not to strongly favor incorporation of miscellaneous toxic ions into the solid. The second reason is that, once incorporation is achieved, it need not be permanent. Many incorporations involve exchange reactions or sorption — and one of the characteristics of ion exchange and sorption reactions is that they are reversible — with the result that, if the pore fluid concentration of a potentially exchangeable constituent should decrease, as could occur, for example, during leaching, the solid tends to restore the equilibria by liberating more of the exchanged ion to the pore fluid. However, a slow, controlled release of toxins may be a realistic goal of immobilization strategies. The third and final reason why such strategies have limited success is that the phase balance in cement systems is not fixed but changes with time, even in the absence of environmentally conditioned chemical exchanges. In response to a slow approach to an internal steady-state pseudoequilibrium, the initially formed hydrate phases change, giving way to thermodynamically more stable (or, at least, less metastable) phases. This is especially apparent in modified cements, where the characteristically slow reaction of blend components such as fly ash or slag will slowly change the bulk composition of the hydrated portions of the paste. Thus, most of the early-formed crystalline phases will not only undergo ion exchange with their surroundings but will also undergo dissolution-reprecipitation processes as the system drifts toward thermodynamic equilibrium. Each toxic waste atom or ion fixed into a host is likely to have the opportunity to repartition itself between aqueous phase and a new host lattice. Thus, the concept of "locking" ions into a crystalline host is not necessarily a realistic expectation for a cement matrix: exchanges will continue to occur at a significant rate even in a matrix unperturbed by environmentally conditioned exchanges. Presumably, the presence of soluble ions (Ca^{2+}, OH^-, etc.) may also cause the precipitation of other solubility-limiting phases which are effective in limiting the release of toxic waste species; the development of Mg-containing double salts in slag blends capable of containing Cr^{3+} has been cited as an example.

Precipitation

Precipitation reactions leading to formation of relatively insoluble species would appear to be the easiest case to treat. Yet, in many respects, these are the most difficult reactions to characterize. This arises for four reasons: (1) inability to distinguish clearly between processes involving mineral precipitation, solid solution, and coprecipitation, (2) a poor database relevant to compound formation, complex stability, and solubility at high pH's, (3) the general complexity of the cement system, with numerous possibilities for interaction between cement and waste components, and (4) the possibility of sequential reactions, perhaps leading to initial formation of a metastable phase, but with its gradual replacement by more stable and hence less-soluble phase(s).

These aspects of toxic species behavior suggest that a global approach to the problems is not at present practicable. Certainly the writer is not in a position to provide a comprehensive summary of all the knowledge which might be applicable. Therefore, my approach to precipitation will be to describe some general principles and to give a few selected examples of single species behavior which have received more in-depth study.

Three general behavior patterns are known which limit the solubility of metals in aqueous solution. Some metals, e.g., sodium and potassium, have solubilities which are little affected by pH. Others are precipitated at higher pH and, once precipitated, remain precipitated at all pH's above the critical threshold: magnesium and calcium furnish examples. The third class of metals comprise the so-called amphoteric elements. These are acid soluble but, as the pH is raised, they undergo an initial precipitation. However, their solubility passes through a minimum and again rises at higher pH. This distinction is widely employed in qualitative analysis to distinguish three broad groups of elements within the periodic table. However, the distinction becomes less clear-cut when applied to cement systems. Consider, for example, aluminum, a classical amphoteric cation. In an aqueous solution, as the pH changes from acid to alkali, Al undergoes reactions of the type

$$Al^{3+}aq \rightleftharpoons Al(OH)_4 \rightleftharpoons Al(OH)_4^-$$

Soluble	Insoluble	Soluble
cationic	neutral	anionic
species	species	species

Solubility passes through a minimum at the point at which the neutral complex is the dominant species.

At the high pH of cement, ~13, these simplistic considerations suggest that Al would be relatively soluble. Yet the experimentally determined solubility of Al in cement pore fluid is extremely low, approximately 1 to 2 ppm. Why should this be so? The answer lies in the chemical nature of the matrix itself. In this instance, the equilibrium between solid $Al(OH)_3$ and aqueous solution is irrelevant because of the complexity of the system; Al preferentially combines with calcium and water to form hydrogarnet or, if SO_4^2 is also present, AF_t. Phase relations in the relevant part of the $CaO-Al_2O_3-SO_3-H_2O$ system are fairly well known, and have been reviewed recently.[14] If we consider for simplicity the sulfate-free $CaO-Al_2O_3-H_2O$ system, the phase having the lowest solubility over a range of Ca:Al ratios is hydrogarnet, C_3AH_6. Figure 1-11 shows experimentally determined solubility curves in this system. The solubility of Al is thus greatly decreased by the presence of Ca. In other words, C_3AH_6 becomes the solubility-limiting phase in the composition range of interest and its solubility affects the Al content of the aqueous phase. In this instance, the nature of the solubility-limiting phase is well known, as is its coexistence with other phases and the locus of the relevant solubility curves. Unfortunately, the

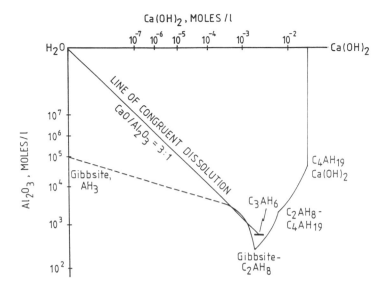

Figure 1-11.   **Representation of some solubility data in the system CaO-A$_2$O$_3$-H$_2$O.**

relevant data for toxic waste species in cement are not always — perhaps almost never — available. Indeed, in the majority of cases, the nature of the solubility-limiting solid, if any, is unknown, and continuing research on toxic waste constituents will be required to determine the occurrence, constitution, and solubility at high pH of such phases.

Leach Tests of Cemented Waste Forms

Leach testing can best be understood against the complex background of cement chemistry and waste-matrix interactions. Leaching in the disposal environment may also remove Ca and the evolution of pH will follow a course similar to that outlined earlier. However, it can be assumed that the alkaline nature of cements will be conserved throughout the lifetime of the materials, at least until the latter stages of dissolution are reached by which time so much material will have been dissolved that the physical integrity of the waste form or barrier will have been impaired.

These conclusions lead to some further remarks concerning simulates for cement pore fluid and leach testing. In an effort to simplify the chemistry of cement, a number of pore fluid simulates have been used in leach testing. Care must be taken to ensure that the simulant is realistic. For example, a saturated Ca(OH)$_2$ solution gives a pH of ~12.4 at 20°C, which corresponds to about 0.025 M, i.e., a relatively dilute solution. Because of its low ion content and lack of buffering capacity, its pH is readily susceptible to change. For example, if this simulate solution is titrated against toxic waste which contains acidic species, the pH of the resulting mix will drop sharply. Moreover, any contamination by atmospheric CO_2 will also rapidly decrease the solution Ca content and lower its pH. In real cements the pH will decrease much less readily: soluble Ca^{2+} and OH^- concentrations are buffered by a solid reserve of $Ca(OH)_2$ or C-S-H, or both, and the alkalinity tends to be maintained.

It is desirable, therefore, to include a solid buffering reserve in simulants. Solid $Ca(OH)_2$ is the simplest buffer but other, more complex, simulants can be devised, for example, of an aqueous NaOH phase buffered with solid C-S-H. Each formulation has merits and demerits but the general principle of maintaining, if possible, a solid alkaline buffering reserve in the course of testing is applicable.

When testing the leach resistance of many materials, e.g., glass, the pH of the leachant is often specified to be near neutral. Cements are, as we have shown, sufficiently soluble as to exert an important conditioning effect on the pH of leachant solutions. Therefore, many conventional leach test protocols which restrict pH to near neutral may not be directly applicable to testing cemented wastes. Broadly, there have been three responses to this. The first is to let the pH rise to its steady-state value on the grounds that this is what will occur in practice, in wet disposal environments. A second type of leach protocol requires that the leachant be renewed periodically, allowing each portion of leachant to equilibrate with cement, in which case pH will be affected by each change of leachant. The third involves maintaining a constant pH — usually close to pH 7 — by automatically adding a noncomplexing acid, such as HNO_3, to the leachant. The first approach generally simulates well those environments in which the ground water regime is static, or nearly so, in which case a high pH aureole will develop around the cemented mass. However, the results are often optimistic with respect to those disposal regimes when more rapid water movement occurs. The second type of test, requiring renewal of leachant, provides a more severe and possibly more realistic assessment of cement performance, especially over the longer term such that ground water flow, advection, and diffusion occur. In this type of simulation it is important to control and record the surface-to-volume ratio of the sample, as well as the renewal schedule and the masses of cement and leachate in order to enable test data to be extrapolated to other conditions, such as those which obtain at disposal sites. The third type of procedure, which maintains a constant pH by adding acid, requires an anion exchanger to prevent high buildup of "inert" ions, e.g., nitrate in the circulating leachant. This procedure measures leaching as well as the acid-buffering capacity of the cement and is thus very severe, perhaps unrealistically so.

It is well known from leaching studies of radwaste species in cement that, when renewal of leachant methods is used, a plot of cumulative fraction leached vs. time characteristically has two distinct portions. An initial rapid leach is followed by a long period of which the cumulative leached fraction is often found to be proportional to $t^{1/2}$. The initial phase, during which accelerated leaching occurs, is often attributed to "wash off". However, the origin of the initial rapid leach is more complex than the term "wash off" might imply. When leachant is first placed in contact with cement, substantial differences between leachant and pore fluid compositions will normally arise. These differences in chemical potentials condition the initial rapid diffusion; a thin leached layer of cement acts as a porous membrane through with chemically activated migration and transport occurs. Normally, many of the membrane pores and channels become blocked as leaching continues; also, differences in ionic potentials become less sharp with the result that a transition to $t^{1/2}$ leach kinetics occurs. The exact mechanisms of early stage leaching are, however, not well understood but, in any event, are reduced in importance if the surface-to-volume ratio is kept low, as will occur in many practical situations.

Because of the slow maturation of cement matrices, especially those made containing slag, fly ash, etc., leach testing of cements should in general not begin before 28 d moist cure, unless special reasons exist for wishing to test immature composites. The writer's experience has been that 60- to 90-d moist cure at 20 to 25°C is desirable prior to commencing leach tests, in order to avoid the problems associated with deconvolution of the short-term data into its separate components, one associated with intrinsic leach performance and the other with the impact of continued cure upon pore structure and, consequently, diffusivity of the matrix.

Case Studies

Several case studies are described involving sorption/lattice incorporation and precipitation of solubility-limiting phases. These are derived from studies on radioactive wastes, but nevertheless the chemical features thus revealed are likely to have strong parallels with toxic but nonradioactive species: in the selected examples radioactivity is not, per se, important with respect to the immobilization potential.

Iodine   Iodine occurs in several speciations in alkaline solutions: Figure 1-12 shows its Pourbaix diagram.[15] The most important speciations at high pH are iodide, I^-, and iodate, IO_3^-; calculation shows that the vapor pressure of elemental iodine, I_2, is extremely low, so vapor-phase transport is not a practicable mode of loss at high pH. The equilibrium between I^- and IO_3^- is attained only slowly in cement matrices; both species have been observed to persist over the timescale of measurement, which now extends over several years.[16] Analysis for speciation has to be made at relatively high concentrations, approximately 10^{-3} to 10^{-4} M, but radioiodine tracers permit total I (all species) to be determined in traces in pore fluids, to about 10^{-8} M. The data show that cement solids are, in general, not good adsorbers for I, although it

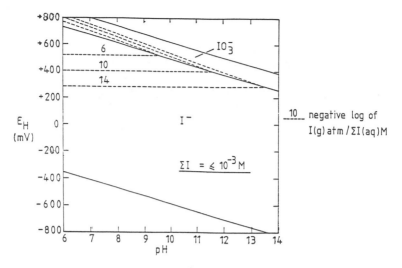

Figure 1-12.   Pourbaix diagram of the iodine system. Superimposed dashed contours and I_2 fugacity.

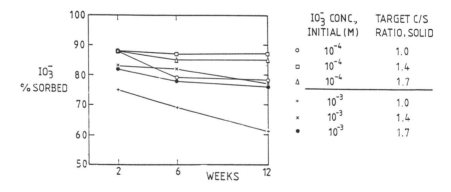

Figure 1-13.   Example of iodine uptake, as IO_3^- on C-S-H. Note that the uptake at ~20°C is affected by the Ca:Si ratio of the gel, IO_3^- concentration, and by time.

is somewhat better incorporated into cements in its IO_3^- speciation than as I^-; Figure 1-13 shows some representative data.[17] The immobilization potential is associated principally with the monosulfate (AF_m) phase in which IO_3^- apparently substitutes for OH^- or SO_4^{2-}. Experiments made using Cl^- to displace sorbed iodine (I^-, IO_3^-) show that iodine species, once incorporated, are not readily displaced by Cl^-. In summary, the behavior of iodine in cement is dependent on speciation. Cement components exhibit weak sorption for both I^- and IO_3^- but crystallochemical inclusion, preferably of iodate, in AF_m-type phases, is also important. High levels of OH^- and SO_4^{2-} prevail in cements, and these compete with iodine for structural sites, with the result that uptake of IO_3^- is modest but significant.

Since the chemical immobilization potential of cement for iodine is rather poor, one can either depend largely on the physical properties of cement for its containment or else seek to find specific "getters" to enhance the immobilization potential of cement. Several investigators have examined solubility product data for various insoluble phases and concluded that Ag offers promise as a specific getter: for iodine immobilization, both AgI and $AgIO_3$ have extremely low solubility products and, moreover, appear to be stable at high pH's. However, when the behavior of silver as a selective precipitant has been reinvestigated,[18] it was noted that previous investigators had included pH, but not E_h, within the scope of their assessment. When this was done, $AgIO_3$ appeared to be of marginal stability in OPC, with AgI showing more promise as a stable phase at E_h values typical of cement. However, when low E_h matrices such as slag cements were used or reducing conditions superimposed from other sources, direct experimental evidence was obtained that AgI slowly decomposed. The mechanisms was not completely elucidated, but for slag cements it was surmised that Ag^+ in AgI was being reduced to Ag^0, liberating iodide, with some resolubilization of silver occurring as its thiosulfate complex. Thus, immobilization by silver was deemed to be impracticable if the cement matrix had a low E_h and a reduced S chemistry; the combination of low E_h with a chemistry of reduced S was inimical to the stability of either or both AgI and $AgIO_3$.

This case study leads to a general conclusion: preliminary assessment of the problem, with a view to its formation in thermodynamic terms, must not be too simplistic. In a complex system, the many possible interactions require careful assessment in order to produce a valid model which will correctly predict the immobilization potential.

Uranium   The oxidation state of uranium in alkaline solutions seem to be limited to U^{6+} and U^{4+}, although the E_h-pH stability regime of the tetravalent species is uncertain at high pH. Certainly the U^{6+} state appears to be stable in Portland cement. $UO_3 \cdot nH_2O$ itself is slightly soluble: it does, however, decrease in solubility in basic solutions; the initial product of precipitation is a nearly amorphous hydroxide of uncertain constitution. It is not known how rapidly this hydrous oxide precipitate will crystallize upon aging. Some predictions about the solubility of U have been made, assuming crystalline $UO_3 \cdot nH_2O$ to be the solubility-limiting phase and using solubility data from standard compilations.[19] However, the underlying assumptions are questionable; in cement systems where U^{6+} or its freshly precipitated species are in contact with both Ca and Si, a series of phases develop which include cement components and which define the solubility properties. Thus, at longer ages, the solubility and crystallinity of hydrous uranium oxides may not be relevant. Two phases

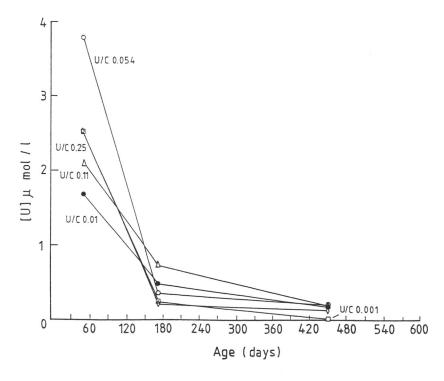

Figure 1-14.   Trends in U solubility at 20°C as a function of U:Ca (U/C) ratio and time. Solubilities at <30 d are high and lie off the scale.

appear to be important with respect to controlling solubility; one is isostructural with the naturally occurring mineral uranophane, the other is close to $Ca_3UO_6 \cdot nH_2O$ where n ~1.5: this latter phase does not appear to have any natural analogue. Uranium-doped cements have been monitored over a period of several years; results of U solubility in the pore fluid are shown in Figure 1-14.[20] The initial rather high solubility of U (not shown) has tended to decline by several orders of magnitude over the period of observation, which at this writing extends to several years. Experiments in which synthetic uranophane is shaken with water conditioned to a high pH suggests that the trend to lower solubilities would be expected, as these experiments also yield low solubilities, within the 10^{-1}- to 10^{-2}-μM range. The important point is that it may take some time, perhaps several years at ~20°C, to attain steady-state solubilities which are indicative of the true immobilization potential of the cement matrix: this is because full development of the solubility-limiting phase occurs only slowly. Hexavalent uranium appears to be a case where sorption and lattice inclusion in cement phases are slight; instead, even at low concentrations, specific mineral precipitation occurs. The solubility trends are dominated in the short

Table 1-4. Precipitation of Waste Species Induced by Cement

Specific Nature	Example	Comment
Precipitation by OH^-	Fe^{3+}, as $Fe(OH)_3$	Rapid
Precipitation by soluble silicate	Amorphous hydrated silicates of Fe, Ni, Cu, etc.	Rapid
Reaction with cement constituents to form solids, e.g., basic salts	Pyroaurite structured compounds related to "green rust"	Speed of formation variable: may be slow

term by pH-conditioned precipitation of hydrous oxides but, in the longer term, by the much lower solubilities of the hydrous compounds of uranium with calcium and silicon.

General Considerations

These considerations, including the case studies cited, suggest that it may be necessary to undertake focused basic research in order to establish the immobilization potential of cement systems for selected toxic species. Furthermore, short-term tests might not be acceptable as predictive of long-term performance. In the examples given, solubility is often found to decrease progressively with time. Probably this type of behavior will be more generally observed than the converse; as rapidly precipitated or sorbed materials tend to crystallize and to react with cement compounds, a drift toward lower free energy and, consequently, toward lowered solubility will occur. However, as the amorphous cement phases gradually improve in long-range order or crystallize, species which are preferentially occluded but which are incapable of participating in mineral precipitation could increase in solubility.

The principal type of solubility-limiting phase which develops spontaneously appears to be examples of so-called basic salt. We are in the process of reviewing basic salt chemistry and structures, with a view to identifying candidate compounds for isolation, synthesis, and solubility measurement. Characterization of those structures which are stable in cements and have significance as hosts for toxics must, of course, remain a long-term research objective.

However, it is certain that the solubility of toxic species and hence the release rate will be very significantly affected by chemical reaction occurring between components of the waste and of cement, and that these will be a factor in controlling the release rate of toxic ions. Table 1-4 summarizes the general nature of the reactions which can be expected. Until the solubility-controlling phases are characterized, it is doubtful if fundamental explanations of the time dependence of immobilization properties and of leach behavior can be achieved; in the immediate future and in the absence of more quantitative data we will have to be content with empirical determination of leach rates. If accelerated tests are employed, they must at least take note of some of the potential pitfalls inherent in the various test methods.

Future Performance of Cement

In the previous sections, we have noted the chemical and physical complexity of cement systems and suggested that, in the initial stages, an account of their immobilization potential should be restricted to a closed-system approach. However, once cements have gone into the intended disposal site, this approach will no longer be tenable. We assume, however, that the disposal environment will remain fairly constant, and the waste form will not be subject to some types of damage which characteristically limit the durability of above-ground concrete, for example, that freeze-thaw, impact loading, abrasion, etc. will be relatively unimportant in disposal scenarios. Nevertheless, cements will certainly react with their local environment. We can distinguish two types of reactions which may affect future performance: intrinsic and extrinsic. Intrinsic reactions are those which are conditioned within the cement matrices, as by reaction between cement and waste components, while extrinsic reactions are those conditioned by the local environment, and embrace reactions occurring with ground water, with local minerals, etc. Probably the best which can be done, considering the present state of knowledge, is to provide a summary of possible reactions in these categories and describe briefly the probable impact of these factors on the performance of cements; Table 1-5 and Figure 1-15 provide the focus for discussion.

The intrinsic matrix stability of cement matrices is good. Experience of historic concrete, up to approximately 150 years for Portland cement, does not disclose any intrinsic propensity for deterioration and the C-S-H gel phase, responsible for pH buffering, is persistent. Numerous examples of Roman pozzolanic cement, not much different in bulk composition from modern fly ash blends, are found to remain coherent in the natural environment. Large masses of low-strength Roman concretes are often subject to settlement cracking, which increases their geometric surface area for leaching, but otherwise they remain physically intact and coherent.

It is probable that the consequence of waste-matrix interactions will prove more difficult to analyze, especially as no historic analogues exist. As noted previously, many substances interfere with strength development of cement; although strength per se is not an important requirement, failure to attain reasonable strength often serves as an indicator of poor microstructure. Open, porous microstructures are undesirable as they lead to poor leach resistance. Moreover, waste-matrix interactions may lead to other, largely unforeseen

**Table 1-5. Factors Affecting the Future Performance of
Cement Matrices**

Intrinsic Factors (Closed System)

- Matrix alteration: mineralogical changes leading to lower solubility,
  changes in pore structure and permeability
- Matrix stability with waste: gas generation; generation of acids,
  complexing agents, colloids
- Cracking owing to shrink-swell, gas generation, etc.

Extrinsic Factors (Open System)

- Dissolution by ground water components leading to pH neutralization or
  expansion/contraction
- Ion exchange with soil/rock components
- Wet-dry cycling
- Microbiological deterioration

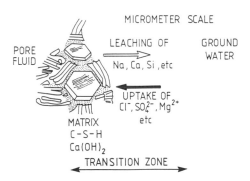

Figure 1-15.   **Pictorial view of some exchanges between cement and
ground water.**

consequences. For example, organics can decompose yielding complexing
agents which may enhance cement and/or toxic ion solubility: colloids may
form which migrate readily; acids may develop which subsequently attack the
matrix and lower its pH, or gas generation may occur which could lead to
unsustainable internal pressures.

Extrinsic factors are, however, likely to cause the most immediate problems.
While cements are not normally prone to microbiological attack, lowering of
the pH, coupled with the presence of organic nutrients, could present potential
problems. Other factors relate to ground water attack. These are largely site
specific, and it is impossible to describe every aggressive environment which

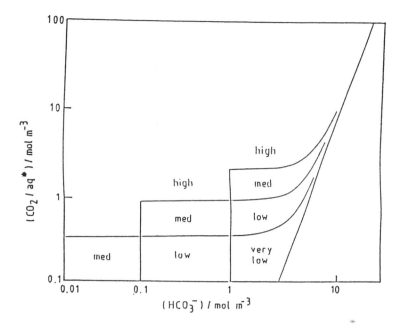

Figure 1-16.   **A classification of the aggressiveness of natural waters containing CO_2 toward cement.**

may be encountered. The simplest possibility, that of straightforward dissolution, has been the subject of mathematical modeling by several groups, and dissolution models are well developed.[21,22] The impact of ground water may, however, result in the ingress of other components. In this respect the local ground water composition is of paramount consideration. Very pure ground waters containing dissolved CO_2 are often surprisingly aggressive toward cement: Figure 1-16 (from Reference 23) depicts one type of classification whose aim is to predict the dissolution power of a ground water. The dissolution power is related to ground water pH, carbon dioxide content, and the CO_2 speciation. More strongly mineralized ground waters, especially those containing Mg and SO_4, will also prove disruptive to cements but for different reasons: Mg will exchange for Ca in $Ca(OH)_2$, while high SO_4^{2-} concentrations develop progressively more ettringite. The combined impact of these two processes leads to expansive cracking and loss of physical coherence of cement, and hence a large increase in geometric surface area available for leaching, as well as to a decrease in pH within the affected portion. Wet-dry cycling may also be a problem; each wet cycle sucks fresh ground water into the pores and each dry cycle precipitates solids which tend to disrupt the matrix. Wet cycles effectively renew the leachant, and the cycle can be continued. Thus, wet-dry cycling generally accelerates the leach process. Many soil minerals also exhibit ion exchange characteristics and, in regimes of low ground water flow, may remove OH^-; this may prevent or handicap development of a protective envelope of cement-conditioned, high pH ground water in the immediate vicinity of the cemented waste. Many of the processes discussed have received at least limited quantification with respect to the basic mechanisms involved, but other aspects of these processes can only be quantified on a site-specific basis.

Disposal strategies will probably have as their objective a slow, controlled release of toxics. The present indications are that this can be achieved, at least in principle, by appropriate design of cement-based systems and of appropriate siting for disposal. There are still research needs, some of which have been indicated in the course of this brief account. In particular, the intrinsic matrix reactions between cement and waste species can usefully form the focus of more effort in order to define long-term solubility properties and release rates.

This account of future performance assessment perhaps fails adequately to highlight the importance of computer modeling. The large number of chemically active system components has been emphasized, as well as the importance of the physical quality of the waste form, its disposal environment, and the "openness" of the system. It is the writer's opinion that the large number of interactions between variables can best be assessed and manipulated in an interactive mode, using a computer. Basic codes to do this already exist or can readily be adapted to this purpose. However, what cannot be readily provided is the database. Operators also require sufficient knowledge of the problems so as to enable the right questions to be asked and finally need sufficient knowledge to assert with confidence that the solutions which emerge are intelligible and reasonable in the context of our present knowledge. These, coupled with acquisitions to the database, are seen as essential to verification and acceptability of cement-based waste-conditioning processes.

ACKNOWLEDGMENT

The writer has benefited from discussion with his colleagues Drs. Atkins, Kindness, and Sagoe-Crentsil which have led to improvements in the interpretation and presentation. He also thanks Dr. E. Lachowski for Figure 1-6.

REFERENCES

1. Glasser, F. P., D. E. Macphee, and E. E. Lachowski. 1987. "Solubility Modelling of Cements: Implications for Radioactive Waste Immobilization" in *Scientific Basis for Nuclear Waste Management X*. Bates, J. K. and W. B. Seefeld, Eds. Materials Research Society, Pittsburgh. pp. 331–341.
2. Brown, P. W. 1989. "Phase Equilibria and Cement Hydration" in *Materials Science of Concrete I*. Skalny, J. P., Ed. American Ceramic Society, Inc., Columbus. pp. 73–94.

3.   Bentz, D. P. and E. J. Garboczi. 1991. "Percolation of Phases in a Three-Dimensional Cement Paste Microstructural Model." *Cement Concr. Res.* 21: 325–344.

4.   Taylor, H. F. W. 1990. *Cement Chemistry.* Academic Press, London.

5.   Garboczi, E. J. and D. P. Bentz. 1991. "Digitized Simulation of Mercury Intrusion Porosimetry" in *Advances in Cementitious Materials,* Vol. 16. American Ceramic Society, Columbus, OH. pp. 365–379.

6.   Atkinson, A., J. A. Hearne, and C. F. Knights. 1990. "Thermodynamic Modelling and Aqueous Chemistry in the $CaO-Al_2O_3-SiO_2-H_2O$ System." Department of the Environment, London. (Report DOE/RW/90/033).

7.   Macphee, D. E., E. E. Lachowski, and F. P. Glasser. 1987. "Compositional Model for C-S-H Gels, their Solubilities and Free Energy of Formation." *J. Am. Ceramic Soc.* 70(7): 481–485.

8.   Macphee, D. E., K. Luke, F. P. Glasser, and E. E. Lachowski. 1989. "Solubility and Ageing of Calcium Silicate Hydrates in Alkaline Solution at 25°C." *J. Am. Ceramic Soc.* 72(4): 646–654.

9.   Macphee, D. E., M. Atkins, and F. P. Glasser. 1989. "Phase Development and Pore Fluid Chemistry in Ageing Blast Furnace Slag-Portland Cement Blends" in *Scientific Basic for Nuclear Waste Management XII.* Lutze, W. and R. D. Ewing, Eds. Materials Research Society, Pittsburgh. pp. 475–480.

10.   Atkins, M., J. Cowie, F. P. Glasser, T. Jappy, A. Kindness, and C. Pointer. 1990. *Scientific Basis for Nuclear Waste Management XIII.* Oversby, V. M. and P. W. Brown, Eds. Materials Research Society, Pittsburgh. pp. 117–127.

11.   Naish, C. P., P. H. Balkwill, T. M. O'Brien, K. J. Taylor, and G. P. Marsh. 1990. "The Anaerobic Corrosion of Carbon Steel in Concrete." AEA Decommissioning and Radwaste, Harwell, U.K. (AEA-D&R-0108).

12.   Bidoglio, G., A. Avogadro, and A. De Plano. 1985. "Influence of Redox Environments on the Geochemical Behaviour of Radionuclides" in *Scientific Basis for Nuclear Waste Management IX.* Werme, L. O., Ed. Materials Research Society, Pittsburgh. pp. 709–716.

13.   Miyata, S. 1983. "Anion Exchange Properties of Hydrotalcite-Like Compounds." *Clays and Clay Minerals* 31: 305–311.

14.   Atkins, M., F. P. Glasser, A. Kindness, and D. E. Macphee. 1991. "Solubility Data for Cement Hydrate Phases (25°C)." Department of the Environment, London. (DOE/HMIP/RR/91/032).

15.   Pourbaix, M. 1966. *Atlas of Electrochemical Equilibria in Aqueous Solution.* Pergamon Press, Oxford.

16.   Atkins, M. and F. P. Glasser. 1990. "Encapsulation of Radioiodine in Cementitious Waste Forms" in *Scientific Basis for Nuclear Waste Management XIII.* Oversby, V. N. and P. W. Brown, Eds. Materials Research Society, Pittsburgh. pp. 15–22.

17. Atkins, M., A. Kindness, F. P. Glasser, and I. Gibson. 1990. "The Use of Silver as a Selective Precipitant for ^{129}I in Radioactive Waste Management." *Waste Management* 10: 303–308.

18. Atkins, M. and F. P. Glasser. in press. "Application of Portland Cement-Based Materials to Radioactive Waste Immobilization." *Waste Management.*

19. Waner, H. and I. Forest. in press. "Chemical Thermodynamics of Uranium." Nuclear Energy Agency, OECD Gif-sur-Yvette, France.

20. Atkins, M., F. P. Glasser, and L. Moroni. 1991. "The Long Term Properties of Cement and Concrete" in *Scientific Basis for Nuclear Waste Management XIV*. Abragano, T. and L. H. Johnson, Eds. Materials Research Society, Pittsburgh. pp. 373–386.

21. Lundén, I. and K. Anderson. 1991. "Interaction of Concrete and Synthetic Granitic Groundwater in Air and in Nitrogen Atmosphere." *Radiochem. Acta* 52/53: 17–21.

22. Cowie, J. 1991. "Modelling Cement-Groundwater Interactions." M.Sc. thesis, University of Aberdeen.

23. Pisters, H. 1963. "Reaction of Corrosive Water with Cement." *Vom Wasser* 30: 208–221.

2 CHEMISTRY OF CEMENTITIOUS SOLIDIFIED/STABILIZED WASTE FORMS

J. R. Conner

2.1. INTRODUCTION

Any discussion of the chemistry of solidified/stabilized (S/S)* waste forms must consider three basic elements: the waste, the S/S reagents, and the environment in which the waste form will exist. This gives rise to three chemical interaction subsets which are useful ways of viewing and approaching the very complex overall chemistry of most S/S systems:

1. The system as an adulterated, weak "concrete", with the waste providing both water (in most cases) and "aggregate". Additional aggregate may be supplied by bulk additives such as flyash; minor S/S additives and reactive impurities in the waste are viewed in the same way as the deliberate additives sometimes used in making concrete.
2. The system as a treated waste, with the S/S reagents being merely treatment chemicals in the same sense as in the wastewater field.
3. The system in which the waste-S/S reagent combination interacts with its disposal or end-use environment.

* The term "solidification/stabilization", abbreviated as S/S, will be used throughout this paper in keeping with standard EPA terminology.

0-87371-748-1/93/$0.00+.50
© 1993 by Lewis Publishers

While the last subset is usually the one of ultimate interest, its complexities can be more easily dealt with by achieving some understanding of the other two approaches. Therefore, these will be examined, at least in concept, before we discuss the area of environmental interaction.

2.2. THE WASTE FORM AS A "CONCRETE"

S/S, indeed the whole environmental field, has been dominated since its inception by the engineering disciplines. These tend, in turn, to stress the physical aspects of waste treatment. As a result, many investigators have viewed S/S as a physical containment process with the attendant emphasis on properties such as compressive strength, permeability, and resistance to weathering. This has been especially true in the nuclear waste field, where reliance on monolithic waste forms is standard procedure, and where nonhazardous liquid wastes and sludges are being solidified. As we shall see, however, physical properties per se are not very meaningful as performance indicators for hazardous wastes disposed of in landfills under the U.S. Environmental Protection Agency's (EPA) regulations promulgated under the Resource Conservation and Recovery Act (RCRA). With this in mind, let us look at cement-based S/S processes from the viewpoint of S/S reagents and their chemistry.

General Chemistry of Portland Cement

Different S/S processes exhibit different setting and curing reactions — the chemical reactions during which the properties of the final waste form are developed. Most, however, solidify by very similar reactions, which have been thoroughly studied in connection with Portland cement technology. While the pozzolanic reactions of the processes using flyash and kiln dusts are not identical to those of Portland cement, the general reactions are alike. One reason for this is presented in an interesting way by Cote.[1] The compositions of most of the primary reagents used in inorganic S/S systems were plotted on a ternary diagram using the three oxide combinations, SiO_2, $CaO+MgO$, and $Al_2O_3+Fe_2O_2$. All of these reagents have the same active ingredients as far as solidification reactions are concerned; this is shown in Figure 2-1.[2] The combinations of these five oxides express the essential composition of any of these materials.

The Portland cement (or simply "cement") which we know today is made by heating together limestone and clay (or some other source of silica) at about 2700°F, forming a mass called clinker. A small amount of gypsum is added and the clinker is ground to a fine powder. The ASTM provides for eight types of Portland cement, while the Canadian Standards Association (CSA) provides for five. The cements of primary interest in S/S, however, are

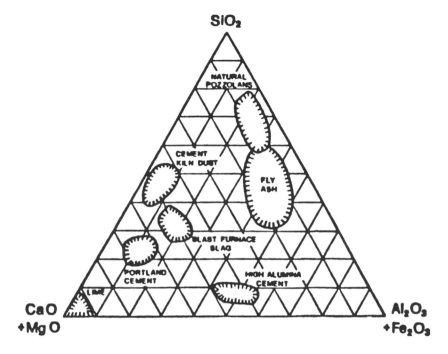

Figure 2-1.   Ternary phase diagram showing common S/S reagent compositions. (From Conner, J. R. 1990. *Chemical Fixation and Solidification of Hazardous Wastes.* Van Nostrand Reinhold, New York. pp. 293–298. With permission.)

- ASTM type I — General purpose Portland cement, and usually the least expensive.
- ASTM type II — General use where moderate sulfate attack is expected, or where moderate heat of hydration is required (CSA "Moderate").

Portland cement is basically a calcium silicate mixture containing predominantly tricalcium and dicalcium silicates, which in cement shorthand are called C_3S and C_2S, respectively,* with smaller amounts of tricalcium aluminate and a calcium aluminoferrite with the approximate formulae C_3A and C_4AF, respectively.[3] Typical weight proportions in an ordinary type I cement are 50% C_3S, 25% C_2S, 10% C_3A, 10% C_4AF, and 5% other oxides.

With a cement S/S process, water in the waste reacts chemically with Portland cement to form hydrated silicate and aluminate compounds. Solids in the waste act as an aggregate to form a "concrete," although the types of solids

---

* This notation system represents calcium, silicon aluminum, and iron oxides by C, S, A, and F, respectively. The subscripts denote the relative mole ratios of each component, e.g., $2CaO \cdot SiO_2$ is C_2S.

encountered in wet wastes usually produce a concrete of low strength. The optimum combination of waste and cement, the type of cement chosen, and any additives will vary with the waste type and its composition. Cement requires a minimum amount of water to obtain workability. This minimum water to cement ratio is approximately 0.40 by weight, but will depend on the waste itself, since some waste solids may absorb large amounts of water. The addition of too much water may result in a layer of free-standing water on the surface of the solidified product, as well as reduction in strength and increase in permeability of the final product.

There is still a surprising amount of disagreement among scientists in this field about how cement combines with water. Two models, the crystalline and the osmotic or gel, emphasize different mechanisms.[4] The former is more widely accepted in the U.S.; the latter was developed by Double and Hellawell[5] at Oxford, England. In both models, the same basic reactions occur. In the presence of water, each of the major crystalline compounds hydrates, but the products are different and their contributions to the final waste form are different. Tricalcium aluminate and sulfates react almost immediately to form hydrates. If sufficient sulfate is present, the reaction product is hydrated calcium aluminate sulfate which coats the surfaces of the particles, preventing rapid further hydration. This is why gypsum retards setting. If no gypsum is present, calcium aluminum hydrates form immediately and the system sets. The overall contribution of aluminate and ferrite hydrates to the strength of the cement paste is thought to be minor.[6]

Strength development after setting is primarily due to C_3S and β-C_2S, both of which give the same reaction products — $Ca(OH)_2$ and calcium silicate hydrate gel (CSH). All of the reaction products, with the exception of lime, have low solubilities in the lime-saturated medium of the cement paste compared to the anhydrous minerals in the original cement. The basic hydration reactions[6] are given in Table 2-1. Reaction starts when the cement powder and water are mixed together. C_3A hydrates first, causing the rapid setting which produces a rigid structure. Setting rate is controlled by the amount of gypsum added in the cement manufacture. The ettringite which forms does not contribute to setting, but coats the cement particles and retards the setting reactions. If a large amount of gypsum is present, secondary gypsum precipitation results, causing quick setting. Hydration of C_3S and C_2S, which accounts for approximately 75% of the cement by weight, is responsible for strength development after the initial set.

Both models end in the same result — the cement grains become interlocked, first setting the cement and finally hardening it. There are, however, still some significant unsolved problems in Portland cement chemistry:[6] the role and effect of minor components; the exact mechanisms of hydration of the cement minerals in the presence of other components and admixtures; the exact structure of the cement paste and its effect on the engineering properties of waste forms; how the properties of cement-based waste forms can be im-

Table 2-1. Basic Hydration Reactions

Reactants	Products	Heat Evolved (cal/g)
$C_3A + 6H$	C_3AH_6	207
$C_3A + 3CS\tilde{} + 32H$	$C_6AS\tilde{}_3H_{32}$ (ettringite)	347
$2C_3S + 6H$	$C_3S_2H_3 + 3CH$ (tobermorite gel)	12
$2C_2S + 4H$	$C_3S_2H_3 + CH$ (tobermorite gel)	62
$C + H$	CH	279

Note: $C = CaO$, $A = Al_2O_3$, $S = SiO_2$, $H = H_2O$, $S\tilde{} = SO_4$.

Adapted from Skalny, J. and K. E. Daugherty. 1972. *Chemtech.* 38–45.

proved. For example, in the latter case, modification of the microstructure, which has been so successful in metallurgy, has not been used with cement.

Several major factors affect the morphology and performance of cement-based waste forms. Fineness affects the rate of hydration, with finer grinds showing accelerated strength development during the 1st 7 d.[7,8] Considerable heat is liberated by the exothermic hydration reactions. This is generally of no consequence; in fact, it helps to speed up the curing process. However, in massive pours of concrete, or waste forms, heat transfer limitations may cause the development of temperatures that affect the final physical properties of the solid. The rate of heat development is determined by the cement composition, and also by certain waste constituents.

The water-to-cement (W/C) ratio is very important, even though it cannot always be optimized in solidification work. The volume of the cement itself approximately doubles upon hydration,[9] creating a network of very small gel pores. The volume originally occupied by the added water forms a system of much larger capillary pores. As the W/C ratio increases, the percentage of larger pores increases, substantially increasing the permeability of the waste form.[1] In a pure W/C system, the permeability is essentially zero at W/C ratio of 0.32, but increases exponentially as the W/C ratio reaches 0.6 to 0.7. Very high W/C ratios will leave "bleed water", or water that appears as standing water on the surface of the solid mass. This occurs at a W/C ratio of about 0.5 or greater. The addition of gelling agents such as sodium silicate and/or bulking agents such as flyash allow much higher W/C ratios to be achieved without generation of bleed water. They may, however, alter the final physical and chemical properties of the waste form and they do affect set time. The cementitious or pozzolanic reactions require sufficient "free" water if they are to go to completion. The water content of a waste as measured by total solids determination is not necessarily all available for reaction.

The long-term durability of concrete and cement mortars is well proven. Nevertheless, chemical and physical attacks due to environmental conditions do occur. The chemical attacks "are interfacial phenomena which take place through complex mechanisms which include ion exchange, dissolution of hydrated solids or formation of new insoluble compounds."[1] Sulfates are the most destructive compounds normally present in the environment because they react with aluminates in the cement structure to form the expansive sulfoaluminate, ettringite, as well as acids. The former break down the physical structure of the waste form. Acids leach lime from the structure until 10 to 15% of the original weight of the wet cement has dissolved, but this has little effect on the strength of the waste form[10] because the Portland cement matrix will resist attack from solutions with pH as low as 5. Strong acids, however, can completely dissolve the matrix.

In S/S work, the amount of lime produced in cement hydration is of importance. A typical cement composition will produce about 30% $Ca(OH)_2$ (0.3 g per 1.0 g of cement), which corresponds to an acid neutralization capacity of 8 meq/g of dry cement.[1] This produces a pore solution with pH in the range of 12 to 13. The cementitious reactions which occur in these processes require the pH to be above 10. This is initiated by the dissolution of free lime from the solid, and continues throughout the setting and curing stages of the mixture. A "false set" can happen when enough free lime is present initially to start the reactions, but availability decreases as the process proceeds and the reactions stop or slow down. Therefore, any reaction that competes successfully for the calcium ion may inhibit setting.

2.3. THE SYSTEM AS A TREATED WASTE — THE WASTE/CEMENT SYSTEM

When the S/S process is viewed as a waste/cement/water system, with the cement acting as a treatment chemical, the chemistry begins to change from what we have just discussed and the complexity increases manifold. The latter is due to the high degree of complexity and variability of wastes themselves, accompanied by our very limited knowledge of the speciation of metals in wastes. The chemistry of this system, and the subsequent properties of the waste forms created, is determined by both chemical and physical interactions of cement and waste. Here, we will deal primarily with the effects of wastes and waste constituents on solidification, leaving the discussion of immobilization or stabilization or fixation of constituents until we look at the waste/cement/water/environment system.

Physical Factors Affecting Solidification

The following physical characteristics of the waste can affect the setting time, curing time, strength, and other physical and chemical properties of the final waste form. It is beyond the scope of this paper to discuss these in detail, except to list them, but further information can be obtained from Reference 2:

- Particle size and shape
- Free water content
- Solids content
- Specific gravity/density
- Viscosity
- Temperature and humidity

Chemical Factors Affecting Solidification

More important to this discussion are the many factors which retard, inhibit, and accelerate the setting and curing of S/S systems. Some also affect the final strength, permeability, and other physical properties of the fully cured* waste form. Many of the compounds, materials, and factors which are known to have such effects are listed in Table 2-2. These have been accumulated from a variety of specific sources[2,11-20] as well as general knowledge in the field of cement technology. The effects are many and varied and are not simple to predict from knowledge of the composition of the waste. Often, a number of species are present, sometimes with opposing effects. The same species may have opposite effects depending on concentration.

Some of the general effects are interesting. Ion exchange can inhibit or retard S/S reactions by removing calcium from solution, preventing it from entering into the necessary cementitious reactions. It can also accelerate the process by removing interfering metal ions from solution. Which occurs may depend on the selectivity of the ion exchange material. Other metals may retard and inhibit the reactions by substituting for calcium in the cementitious matrix, which may explain the effect of magnesium in dolomitic lime and lime products. Certain substances are natural or synthetic complexing agents which remove calcium from availability in the setting and curing reactions. On the other hand, alcohols, amides, and specific surfactants can aid in wetting solids

---

* The term "fully cured" is relative. Most pozzolanic and cementitious reactions continue for very long periods of time, perhaps forever. The changes in physical properties which take place in the long term are usually small, but may be significant in certain situations. "Fully cured" is usually taken to be periods ranging from 10 to 100 d, with 28 d being the most common.

Table 2-2. Factors Affecting Solidification

Compound or Factor	Effect[a]	Mechanism Affected[b]
Fine particulates	I, P	P
Ion exchange materials	I, A	I
Metal lattice substitution	I, A	I
Gelling agents	R, I, P–	P, I, M
Organics, general	I, P, R	I, D
Acids, acid chlorides	P–	I
Alcohols, glycols	R, P–	I, W
Carbonyls	R	I, D
Chlorinated hydrocarbons	P–, R	I, M
Grease	I, P	P
Lignins	I	C
Oil	I, P	P
Starches	I	C
Sulfonates	R	D
Sugars	I, R	C
Tannins	I	C
Organics, specific		
Ethylene glycol	P	I
Hexachlorobenzene	P–, P+	I
Phenol	P–	I
Trichloroethylene	P–	I
Inorganics, general		
Acids	P–	I
Bases	P–	I
Borates	R	M
Chlorides	R, P	I
Chromium compounds	A	I
Heavy metal salts	P–, A, R	I
Iron compounds	A	F, M
Lead compounds	R	M
Magnesium compounds	R	M
Salts, general	P–, A, R	I
Silicas	R	F
Sodium compounds	I	I
Sulfates	R, P	I
Tin Compounds	R	M
Inorganics, specific		
Calcium chloride	A, R	M
Copper nitrate	P+	I
	P+, P–	I

Table 2-2. Factors Affecting Solidification (continued)

Compound or Factor	Effect[a]	Mechanism Affected[b]
Gypsum, hydrate	R	I
Gypsum, hemi-hydrate	A	I
Lead nitrate	P–, P+	I
Sodium hydroxide	P+, P–	I
Sodium sulfate	P+, P–	I
Zinc nitrate	P+, P–	I

Note: When the effect may be positive or negative, depending on concentration, the first symbol listed represents lower concentration, the last higher concentration.

[a] I = setting/curing Inhibition (long term), A = setting/curing acceleration, R = setting/curing retardation (short term), P+ = alteration of properties of cured product, positive effect, P– = alteration of properties of cured product, negative effect.

[b] P = coats particles, I = interferes with reaction, C = complexing agent, M = disrupts matrix, F = flocculent, D = dispersant, W = wetting agent.

From Conner, J. R. 1990. *Chemical Fixation and Solidification of Hazardous Wastes.* Van Nostrand Reinhold, New York. pp. 293–298. With permission.

and dispersing fine particulates and oil which interfere with reactions by coating the reacting surfaces. Flocculants can also serve this purpose. Some of these techniques are listed in Table 2-3.

Stabilization, as opposed to solidification, involves the effects of the cementitious treatment reagents on the waste, rather than the effects of the waste components on the cement discussed above. This is really the most important aspect of S/S, since it is what affects "human health and the environment", the issue of importance in environmental chemistry. However, stabilization of the waste is only meaningful in terms of interaction of the W/C system with the environment, a process which is studied and measured primarily by means of leaching tests.

2.4. THE WASTE/CEMENT/LEACHATE SYSTEM

The mechanisms of fixation are different for the three primary groups of pollutants: metals, other inorganics, and organics. This paper will confine itself to the fixation or immobilization of metals. Since fixation is generally defined as a treatment product's resistance to leaching in the environment, it is essential to understand the leaching process and the tests by which it is measured.

Table 2-3. Techniques for Countering Inhibition

Method or Material	Mechanism
Flocculent	Aggregation of fine particles and film formers
Wetting agent	Dispersion of oils and greases and fine particulates away from reacting surfaces
pH adjustment	Removal of interfering substances from solution; destruction of gels and film formers
Fe^{+2}/Fe^{+3} addition	Precipitation of interfering substances
Ion exchange	Removal of interfering substances from solution
Sorbent addition	Removal of interfering substances from reacting surfaces
Redox potential	Destruction/conversion of interfering substances
Aeration	Alteration of biological status; removal of interfering volatiles
Temperature adjustment	Acceleration of reaction rate to counter retarding effect
Lime addition	Supplies additional calcium for reaction; reacts with certain interfering organics; pH adjustment
Sodium silicate	Reacts with interfering metals; causes acceleration of initial set
Calcium chloride	Accelerates set in Portland cement systems
Sodium hydroxide	pH adjustment; may solubilize silica for quicker reaction with concentration
Amines, organics	Mechanism unclear or varied
Metal ions	Mechanism unclear or varied

From Conner, J. R. 1990. *Chemical Fixation and Solidification of Hazardous Wastes.* Van Nostrand Reinhold, New York. pp. 293–298. With permission.

What is Leaching?

If ground or surface water contacts or passes through a material, each constituent dissolves at some finite rate. When a waste, treated or not, is exposed to water, a *rate* of dissolution can be measured. We call this process *leaching,* the water with which we start the *leachant,* and the contaminated water which has passed through the waste the *leachate.* The capacity of the waste material to leach is called its *leachability.*

Leaching is a rate phenomenon and our interest, environmentally, is in the *rate* at which hazardous or other undesirable constituents are removed from the waste and into the environment via the leachate. This rate is usually measured and expressed, however, in terms of *concentration* of the constituent in the leachate. This is because concentration determines the constituent's effects on living organisms, especially humans.* Concentration is the primary basis for water quality standards and water quality standards, especially drinking water standards,[21] are normally the basis for leaching standards. Thus, we speak of the leaching rate of a constituent, but usually measure it as concentration in the leachate.

When evaluating a material for leachability, we usually compare the concentration of the hazardous constituent in the leachate to that in the original waste. This tells us what proportion of the constituent dissolved out during the test, which becomes a measure of the leachability of the material.

Measurement of Leachability

For obvious reasons, actual tests cannot be run on the solidified waste in the specific disposal site to determine leachability over thousands of years before choosing a S/S technology. Therefore, we must resort to accelerated tests. It is not possible here to explore the details of leaching test methods and results; a good summary is given by Darcel.[22] The environmental acceptability of a hazardous waste for land disposal in the U.S. is now usually based on the U.S. EPA Toxicity Characteristic Leaching Procedure[23] (TCLP), but other tests may also be used in different regulatory regimes and disposal scenarios.

Most tests are batch procedures in which the waste is contacted with a leachant for a specific period of time, agitating the mixture to achieve continuous mixing. Chemical equilibrium is often obtained,[24] especially when the solidified waste is crushed before extraction. After extraction and separation of the fluid from the solids, the leachate is analyzed for specific constituents. It should be noted that the TCLP uses a leachant-to-waste ratio of 20:1, so that the maximum concentration of constituent which can be attained in the leachate is 5% of that in the original solid. The leachant is dilute acetic acid, buffered in some procedures. The total amount of acid added varies with the test and/or with the alkalinity of the waste. The final pH of the leachate at the end of the test is controlled by the alkalinity of the waste in most cases where the leachant is deionized water or dilute acid. As we will see, final pH is one of the prime controlling factors in metal leaching.

The TCLP test is designed to simulate the leaching potential of a waste in a so-called "mismanagement" scenario, where it is disposed in a landfill

---

* There are also cumulative effects determined on total exposure over a long period of time.

designed for municipal refuse. Such landfills are known to generate organic acids during decomposition of organic matter in the refuse; the purpose of acetic acid in the leachant is to simulate those acids. However, neither the TCLP or most other tests actually simulate any real-world set of conditions. Arguably, however, they may create a worst-case environment for leaching. Attempts to correlate laboratory leaching tests with field data have not been successful.[25] Nevertheless, these procedures are widely used and required as specification tests.

Factors Affecting Leachability

There are two sets of factors, or variables, which affect the leachability of a treated waste: (1) those which originate with the material itself and (2) those which are a function of the leaching test or the disposal environment. The combination of the two sets determines the leachability of the material. Test method factors are discussed in some detail by Conner,[26] and are not further elucidated here. They include surface area of the waste, the nature of the extraction vessel, the agitation technique, the nature of the leachant, the ratio of leachant to waste, the number of elutions used, the time of contact, temperature, pH of the leachant, and the method used to separate extract from solid. Leaching mechanisms, on the other hand, are more fundamentally important because they, in combination with immobilization techniques, determine the chemistry of the system.

Immobilization/Containment Mechanisms

Major waste form factors and immobilization, or containment, mechanisms include

- pH control
- Redox potential control
- Chemical reaction
     Carbonate precipitation
     Sulfide precipitation
     Silicate precipitation
     Ion-specific precipitation
     Complexation
- Adsorption
- Chemisorption
- Passivation
- Ion exchange
- Diadochy
- Reprecipitation

- Encapsulation
    Microencapsulation
    Macroencapsulation
    Embeddment
- Alteration of waste properties

Equilibria   Before discussing specific mechanisms, it is worthwhile reviewing solubility of metal species in terms of the equilibrium constant concept
since so many of the reactions which affect leaching occur between relatively
simple molecules and ions, and are reversible. While this may seem elemen-
tary, it is frequently misunderstood. For any reaction

$$aA + bB = cC + dD$$

the equilibrium constant is expressed as

$$K_e = \frac{[C]^c \times [D]^d}{[A]^a \times [B]^b} \tag{1}$$

where $[A]$, $[B]$, $[C]$, and $[D]$ are the activities of the respective species.

In the case of an ionizable species, such as a metal compound in aqueous
media, most of the compound exists as a solid in contact with the leaching
solution or pore water in the matrix:

$$AB = A + B$$

there, the dissociation constant is expressed as

$$K_d = \frac{[A] \times [B]}{[AB]} \tag{2}$$

The reactions of interest in leaching are heterogeneous, that is, they involve
both a solid and a liquid. In the above equation, the solid is AB, and its
concentration is a constant. Multiplying both sides of Equation 2 by $[AB]$
results in a different value for K_d which we call K_{sp}, the solubility product:

$$K_{sp} = [A] \times [B] \tag{3}$$

From this comes the important conclusion that, for dissociation of solids
such as metal compounds, the concentration of the solid metal compound in the
waste does not affect the concentration of metal ion in the leaching solution.
This remains true as long as any solid metal species exists in contact with the
leaching solution. Furthermore, since K_{sp} is a constant for a given system at a

given temperature, changing the concentration of either A or B automatically changes the concentration of the other, giving rise to the common-ion effect. This is quite important in metal fixation for systems at equilibrium, since the metal concentration in the leachate can be controlled by the addition or removal of its associated anion from another species. It is rarely possible to actually calculate the expected effect with any degree of accuracy in complex waste-reagent systems; however, the degree of departure of measured values from the theoretical concentration indicates the extent of other factors operating in the system.

Much of the confusion about equilibria in S/S systems stems from the fact that there may be several species of a given metal in a waste or stabilized waste form. For example, it is not unusual to find lead existing simultaneously as $Pb(OH)_2$, $PbSO_4$, $PbCL_2$, and perhaps as carbonates and silicates, all in the same waste form. Each species has its own K_{sp} for the medium under consideration — the leachant used in the study or existing in the environment, modified by the dissolved salts from the waste form. Furthermore, the amount of metal in solution is determine by the interrelationship between its various species by the common-ion effect, as well as by common anions in solution and, especially, by pH of the leaching medium.

General Solubility Considerations   The solubilities of various metal species in purewater are reviewed in the literature.[27] However, the differences between simple systems and real ones are illustrated by comparing values for individual species with those obtained in leaching tests on treated wastes. This is done in Table 2-4 for the more common species. Such comparisons assume a certain speciation for the metal, which in most cases has not been confirmed. Nevertheless, the comparison serves a useful function — to point out the limitations of fixation technology in real applications or, in some cases, to show how the complex interactions of several mechanisms can achieve results better than those expected from simple theories.

The aspect of metal fixation which seems most confusing to workers in the field involves the speciation of the metal before S/S treatment. Most wastes that are treated by S/S are not solutions; they are sludges, filter cakes, and other residues from waste water treatment systems. The metals have been precipitated with lime or other alkali, or with other agents such as sulfide, to produce metal hydroxides, sulfides, or other compounds which have low solubility under the conditions of precipitation, usually in the pH range of 6 to 8 and a dilute water medium. In this state, there is little immediate reaction between the metal species and the S/S reagents. If the metal species in the waste is more soluble than the species that would be formed by anions or ligands introduced by the S/S system, there may be gradual respeciation beginning at the particle surface, but total respeciation would be expected to occur only over a long period of time. Furthermore, the soluble anion or ligand may not be available as such for long if it can react rapidly with other components of the S/S system. The ultimate result, then, is usually a mixture of metal species dispersed in a

Table 2-4. Comparison of Published Metal Solubility Values With
Actual Leaching Results on S/S-Treated Wastes

Metal	Published Value (mg/l)	Leaching Test Result[a] (mg/l)
Antimony (S)	1.2	<0.05
Arsenic (S)	0.3	<0.01
Barium (SO$_4$)	1.5	<0.02
Cadmium (S)	3.0×10^{-9}	<0.01
Copper (S)	0.12	<0.002
Chromium (OH)	0.2	<0.01
Lead (OH)	0.001	0.003
Mercury (S)	2.0×10^{-21}	<0.001
Nickel (OH)	6.1	<0.005
Selenium	—	<0.05
Silver (S)	2.0×10^{-12}	<0.003
Zinc (S)	1.0×10^{-4}	<0.005

[a] For waste with significant metal content. Metal Species: (S) sulfide, (OH) hydroxide, (SO$_4$) sulfate.

From Conner, J. R. 1990. *Chemical Fixation and Solidifcation of Hazardous Wastes.* Van Nostrand Reinhold, New York. pp. 293–298. With permission.

cementitious matrix. Simplistic models of waste systems based on a single speciation of a metal in a waste thus are often invalid.

Obviously, it is desirable to know the speciation of the metals in the raw waste. However, in these complex systems, really complete analysis is very time consuming and expensive, requiring specialized equipment and, therefore, is seldom done. The problem of speciation is minimal when the metal species to be fixed is in solution. Unless it is complexed in soluble, stable form, the metal can usually be precipitated from solution as a known species which exhibits minimum solubility under the expected disposal conditions (or in meeting the required leaching test). This is especially important for the heavy metals of environmental interest which exhibit amphoterism, such as arsenic, cadmium, chromium, lead, and zinc. Because most S/S systems are quite alkaline, usually above pH 11 (at least initially), the solubility of the metal hydroxide in the S/S-treated waste may actually be higher than in the original, untreated sludge. This is often the case with Cr^{3+}.

Metals exist in solution in forms other than simple ions or molecules. Actually, metal ions in solution are usually solvated, that is, they are associated with water molecules in a definite arrangement. The maximum number of water molecules associated with the ion depends on the metal's coordination number. When the water molecules are replaced by other ions or molecules, the result is termed a metal complex. The chemical bonds involved are covalent rather than ionic, i.e., electrons are shared by each of the bonded atoms. Common examples of such complexes often encountered in environmental chemistry are the cupriammonium, citrate, gluconate, EDTA (ethylene di-amine tetraacetic acid), and nitrilotriacetate complexes.

Like other chemical species, complexing agents vary in their ability to solubilize metals. Chelate structures exist in solution in an equilibrium mixture with chelating agent (L) and metal ion (M):

$$M^+n + L^-n \rightleftharpoons ML$$

The measure of effectiveness of the chelating agent is the stability constant, K, which is expressed as

$$K_{ML} = \frac{ML}{(M+n)(L^-n)} \tag{4}$$

The larger the positive value of K, the greater is the tendency of the chelate to form. Many of the metals of environmental interest form very stable chelates: cadmium, cobalt, copper, lead, mercury, nickel, and zinc.

As we saw from the discussion on equilibria and Equation 3 above, the total metal content does not, in theory, affect the concentration of metal ion in solution so long as there remains undissolved solid of that metal species in contact with the leachant. It is interesting to compare actual metal leaching with total metal content in real systems. This has been done and the results are shown in Figures 2-2 and 2-3, for chromium and lead, respectively, in a commercial waste type — the very common EPA F006 category, which includes various sludges from wastewater treatment in electroplating pro-cesses. Each of the data points in each graph is from a different, but similar, F006 waste sample, nine in all, which constitute a reasonably homogeneous data set. Each of these wastes was analyzed for total metals, and the TCLP was carried out on both untreated and stabilized samples. The stabilized samples all used the same reagent, a cement kiln dust, at the same mix ratio. They also all had little acid neutralization capacity. Leaching test results for both untreated and stabilized samples were plotted against total metal content for each metal — Cr and Pb.

It is evident from the plots that leaching of metals from untreated wastes shows a direct correlation to metal content, while leaching from stabilized wastes does not (except for what appears to be a small effect in the case of chromium). Plots for nickel and cadmium (not shown here) closely followed the lead results. This makes sense in view of basic solubility concepts. In the case of slightly soluble species, such as metal hydroxides in neutral or alkaline

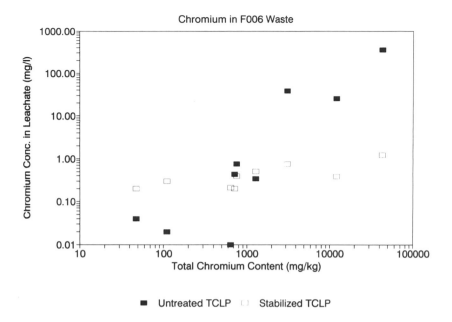

Figure 2-2.   Leaching as a function of chromium content.

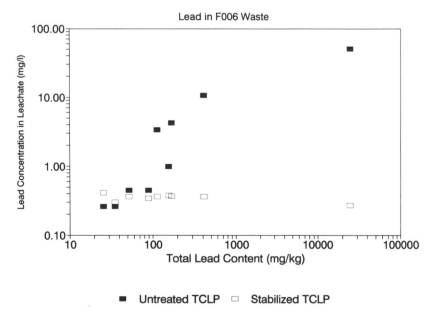

Figure 2-3.   Leaching as a function of lead content.

media (stabilized wastes), solubility should be independent of the amount of solid species present. When the system becomes acidic, as it does in the TCLP test on untreated wastes, hydroxides become very soluble, the solids dissolve, and the amount in solution in the leachate should depend on total metal content, up to the point where metal content exceeds its solubility as metal hydroxide in the medium.

The moderate increase in chromium leaching with total chromium content is probably due to the presence of small amounts of Cr^{6+}, whose concentration is likely to increase with that of total chromium and which is quite soluble even in stabilized wastes. Many chromium wastes come from industrial processes utilizing Cr^{6+} as a chromate or dichromate. The water treatment method used to produce the sludge which is subsequently stabilized usually leaves some Cr^{6+} along with the predominant Cr^{3+} species. Simple S/S systems cannot immobilize Cr^{6+} species, and the hexavalent chromium is leached out. This example of the effect of multiple speciation of the same metal in a single waste is very vivid because of the great difference in solubilities of the two species; with most metals the effect is much more subtle.

Different species may have the same molecular formula, but different crystal structures and/or morphologies, or they may be more or less hydrated versions of the same basic compound. This is discussed later in the section on "Chemical Reaction." In any case, both theory and experience show that metal content has little to do with the required stabilization agent type or mix ratio in slightly soluble systems such as hydroxides, provided that pH during leaching remains in the neutral or alkaline range. Other factors such as speciation of the metal are probably more important, but we have no easy way to determine what species are present.

pH Control   It is widely accepted that cement- and pozzolan-based waste forms rely heavily on pH control for metal containment. In fact, some investigators believe that this is the only important factor in metal fixation. However, more recent work has shown that this is not so, at least when very low leachabilities are required. Using a sequential batch extraction technique, Bishop[28] demonstrated that even after the alkali was leached from cement-based waste forms, lead and chromium leaching was much lower than would be expected from metal hydroxide solubilities. In this case, the metals were probably bound into the silica matrix itself. In other S/S systems, the same mechanism may not be controlling. Nevertheless, pH control is a necessary, if not sufficient, condition for metal fixation in most systems. It has also been shown to affect the leachability of some other inorganic and organic species.

Normally, high pH is desirable because metal hydroxides have minimum solubility in the range of pH 7.5 to 11. However, some metals such as chromium which exhibit amphoteric behavior have high solubility at both low and high pH. Unfortunately, all metals do not reach minimum solubility at the same pH, so that the optimum pH of the system must be a compromise. In addition, the hydroxides of certain metals such as lead exhibit solubilities above regu-

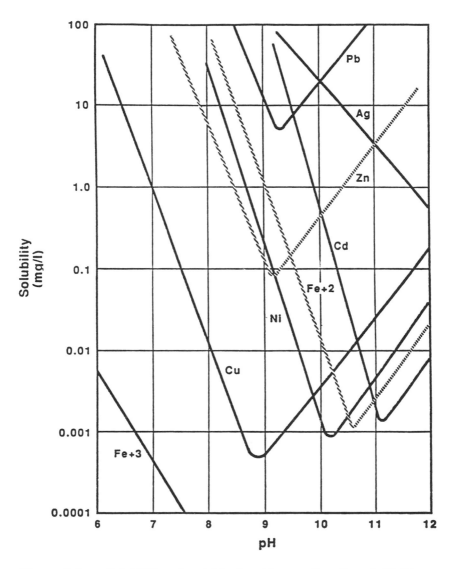

Figure 2-4.   Solubilities of metal hydroxides as a function of pH. (From Conner, J. R. 1990. *Chemical Fixation and Solidification of Hazardous Wastes.* Van Nostrand Reinhold, New York. pp. 293–298. With permission.)

latory limit or goal even at optimum pH. This is illustrated by the solubility curves for various metal hydroxides in water, derived from several sources,[29,30] as shown in Figure 2-4. It is obvious that care must be exercised in using solubility data to predict leachability of any constituent. Such data are obtained experimentally or calculated from stability constants for individual species

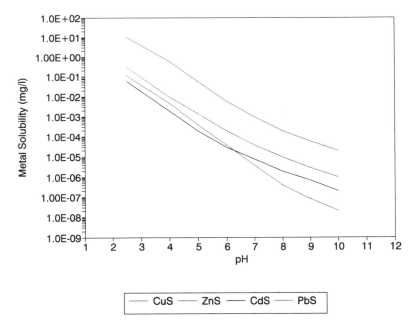

Figure 2-5.   **Solubilities of metal sulfides as a function of pH. (From Conner, J. R. 1990.** *Chemical Fixation and Solidification of Hazardous Wastes.* **Van Nostrand Reinhold, New York. pp. 293–298. With permission.)**

alone, or for relatively simple combinations, in a specific solvent — usually pure water. Depending on the assumptions used in the calculations, or the experimental conditions used, widely different results are obtained. Waste forms, moreover, contain many other soluble species which may change the actual solubility of any metal due to effects such as common ion, complexation, total ionic strength, and redox potential. Certain metals, notably chromium as we have seen, exhibit completely different solubilities in their various valence states, and may exist in cationic or anionic form, or both, in the same waste.

Solubility plots for sulfides are different from those for hydroxides, as shown in Figure 2-5. Their solubilities are much lower than those of the hydroxides under alkaline conditions, with the exception of chromium which does not form a sulfide under these conditions. Furthermore, the sulfides do not exhibit increasing solubilization at high pH, at least not at the pH levels encountered in most S/S systems. Also, the sulfides have adequately low solubilities even at the moderately acid pH levels, pH 4 to 5, used in leaching tests and existing under actual environmental conditions.

In real systems, it is useful to look at the relationship between final leachant pH (at the end of a leaching test) and metal leaching. One approach to S/S formulation has been to pose that pH at the optimum, or minimum, point on the

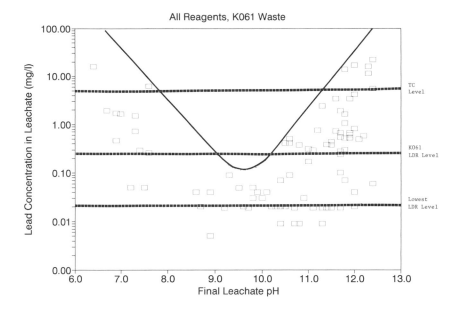

Figure 2-6.   Lead leaching vs. final leachate pH (all reagents).

pH curve that represents lowest leachability of the metal in the system under consideration. This can be done by testing a series of formulations, all using the same reagent system, on the same waste or on a relatively homogeneous group of wastes. One such study was conducted at the author's laboratory on electric arc furnace dust, an EPA-listed (K061) waste of considerable environmental concern. Samples from four different sources were mixed with several different cementitious reagents at various mix ratios. The leachability of the RCRA metals of interest for each of these mixtures was determined by the TCLP test. These simple systems were found to immobilize all metals except lead to RCRA-characteristic (TC) and K061-listed levels; lead exhibited widely varying results that seemed to be related to final leachate pH, so lead leaching was plotted against final leachate pH. The results are shown in Figure 2-6, where the solid line delineates the zone where all the leaching data fall. The dotted lines define various regulatory levels — RCRA toxicity characteristic (TC) and two land disposal restriction (LDR) levels. Thus, a region bounded by a horizontal dotted line and solid lines to the left and right defines the allowable pH range for a particular regulatory leaching requirement. It is evident that any reagent will stabilize lead to the TC level (5.0 mg/l) so long as the pH is between about 7.5 and 11.5. To meet the LDR level for K061, however, the pH must lie in a narrower range — about 9 to 10. Also, it would be impossible to ensure that the lowest LDR level could be met at any pH if, for example, the waste contained other RCRA codes which necessitated that more stringent level.

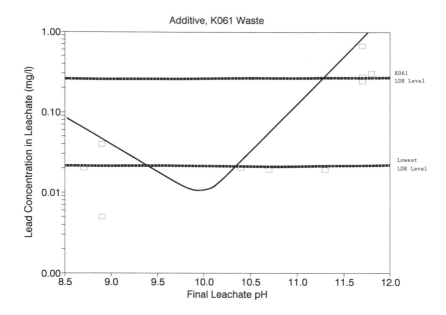

Figure 2-7.   Lead leaching vs. final leachate pH (additive).

To consistently meet the K061 LDR level would necessitate formulating to keep the final leachate pH in the range of 9 to 10, not an easy task when a variety of properties must be achieved in the waste form. Another approach would be to use a reagent formulation that broadens the allowable range, i.e., shifts the curve downward and/or changes its shape. This was done through the use of a proprietary additive in a cementitious system, and the results for ten different formulations on the four different K061 wastes are shown in Figure 2-7. (Here, the shape of the curve is also determined by data not shown in the figure.) Now, the K061 level can be met at any pH between about 8 and 11+, and the lowest LDR level can be met between approximately pH 9.5 and 10.5.

It is important to understand that this is not simply a pH adjustment or buffering effect. Even a precise pH adjustment would only have placed the final leachate pH in the 9 to 10 range, and buffering, by definition, would simply have kept it there. Even if that were satisfactory, in this system it still would not have ensured meeting the lowest LDR level. Also, such precise pH manipulations are often impractical in actual S/S systems and disposal scenarios. The effect of the additive is to respeciate the lead to a less soluble, less pH-sensitive form. How this may happen is discussed later.

Redox Potential   The redox potential, E_h, is the oxidation-reduction potential (ORP) referred to the hydrogen scale, expressed in millivolts. It establishes the ratio of oxidants and reductants existing within the waste-environment system, and may affect the valence state of a metal in that system. The effects of redox processes are commonly expressed as E_h-pH diagrams. The scale ranges from oxidizing (positive) to reducing (negative) potentials, and the resulting diagram shows the domains, or stability fields, in which different species can exist. Such diagrams are useful primarily for conceptual purposes in S/S systems, due to the complexity of the systems and the fact that the boundary lines are regions of transition rather than sharp delineations. Also, the diagram may presuppose that an anionic species shown is, in fact, available in the system. Unfortunately, relatively little attention has been paid to this factor by workers in the field. It is an area which needs much more elucidation, especially as stricter regulatory standards require fixation of multivalent metals to low leachability levels.

Chromium is one example of the influence of valence state on solubility. Arsenic is another. The presence of strong oxidants or reductants can change the valence state of a number of the metals, affecting their chemical speciation and, therefore, their mobilities by orders of magnitude. For some metals such as arsenic, both valence and speciation as either cation or anion can change easily with redox potential.[31] Of the metals of interest in non-nuclear S/S, seven have more than one possible valence state in aqueous systems (As, Cr, Fe, Hg, Mn, Ni, and Se). Also, nitrogen and sulfur have multiple valence states which affect speciation of the metals in a given system; Ag, Cu, Cd, and Zn can be strongly influenced by redox processes even though they have only one valence state in aqueous systems.[32]

A common treatment problem posed by real wastes occurs when oxidizable or reducible species other than the one of interest are present in a waste. An example is the alkaline chlorination destruction of cyanide in electroplating wastes. This treatment will also oxidize the Cr^{3+} present in most such wastes, requiring subsequent reductive treatment. The chromium also consumes more hypochlorite than would be necessary for cyanide destruction, as does the presence of many organics, ferrous iron, etc. Similarly, the use of reducing agents to convert Cr^{6+} to Cr^{3+} can become very expensive if other reducible species are present. Furthermore, in both cases, nontoxic organics may be converted into toxic compounds — a potentially widespread problem which has not been given adequate attention in the past. Chromium is also inadvertently transformed into a soluble species (see previous discussion under "pH Control") by oxidation under high pH conditions.

Arsenic is especially facile in valence state transformations. In this case, the reduced form, As^{3+}, is usually the more soluble, but this varies also with speciation. The solubilization of arsenic under changed redox conditions in the environment is well known.[31] However, the use of oxidation coupled with deliberate speciation can also be used to immobilize arsenic in a stable form. Sandesara[33] describes one such reaction:

$$4As(ONa)_3 + 5FeSO_4 + 4Ca(OH)_2 + 5H_2SO_4 \rightarrow$$
$$FeAs_2 + 2FeAsO_4 + 6Na_2SO_4 + 2Fe(OH)_3 + 4CaSO_4 + 6H_2O$$

Table 2-5. Comparison of Hydroxide and Sulfide Solubilities

	Approximate Solubility (mg/l)		
Metal	Hydroxide	Sulfide	Difference Factor
Iron	5×10^1	1×10^{-4}	5×10^3
Cadmium	3×10^0	1×10^{-8}	3×10^8
Chromium	1×10^{-3}	None	None
Copper	2×10^{-2}	2×10^{-13}	1×10^{11}
Lead	2×10	6×10^{-9}	3×10^8
Mercury	6×10^{-4}	1×10^{-21}	6×10^{17}
Nickel	7×10^{-1}	6×10^{-8}	1×10^7
Silver	2×10	4×10^{-12}	5×10^{12}
Zinc	3×10^2	1×10^{-6}	3×10^8

$$\text{Difference factor} = \frac{\text{Hydroxide solubility}}{\text{Sulfide solubility}}$$

From Conner, J. R. 1990. *Chemical Fixation and Solidification of Hazarouds Wastes.* Van Nostrand Reinhold, New York. pp. 293–298. With permission.

Chemical Reaction   We have already introduced one chemical reaction — hydroxide precipitation. While this is undoubtedly the most common for metal fixation, others are also significant. Metals can be precipitated as carbonates, sulfides, silicates, sulfates, and other simple species, and also as complexes. It has been stated that, in the case of fixation of metals by soils, the reactions are preceded by surface adsorption followed by dehydration and rearrangement of the solid phase.[32] Undoubtedly, many S/S fixation reactions occur in several stages, involving two or more of the mechanisms described earlier. The most important classes of metal compounds in S/S, other than hydroxides, are sulfides, silicates, and carbonates. Others — phosphates, inorganic and organic complexes, and various reactions which bond metals to insoluble substrates — are also occasionally used.

Other than precipitation as hydroxides, sulfide precipitation has probably been the most widely used method to remove metals from wastewater, and so was quickly adopted in S/S treatment. Most metal sulfides are less soluble than the hydroxides at alkaline pH (arsenic is an exception), as illustrated in Table 2-5. The low solubilities required for highly toxic metals such as mercury are often achievable only by speciation as sulfides, since the metal sulfides have solubilities several orders of magnitude lower than the hydroxides throughout the pH range. Also, their solubilities are not as sensitive to changes in pH. However, metal sulfides can resolubilize in an oxidizing environment,[31] and there is currently disagreement over the acceptability of metal sulfide sludges in uncontrolled landfills. It is necessary to maintain pH 8 or above to completely prevent evolution of H_2S. While excess sulfide ion is necessary for the precipitation reaction, the excess must be kept to a minimum so that free sulfide removal treatment is not required before the waste can be landfilled. Precipitation is normally done with Na_2S or NaHS. Another method is the use of very low-solubility species such as FeS or elemental sulfur, which eliminates the free sulfide problem. One interesting exception in sulfide precipitation is that of chromium, which does not precipitate as the sulfide, but as the hydroxide. In some situations, organo-sulfur compounds may have certain advantages over the inorganics.

Another important method of immobilizing metals by chemical reaction is silicate precipitation by the use of soluble silicates. The reactions of polyvalent metal salts in solution with soluble silicates have been studied extensively over many years.[34] The best known process uses a combination of Portland cement and sodium silicate. The "insoluble" precipitates which result from such interactions are usually not well characterized, especially in the complex systems representative of most wastes. Metal silicates are nonstoichiometric compounds in which the metal is coordinated to silanol groups, $SiOH^{3-}$, in an amorphous silica matrix. The reactions of soluble silicates in solution were best summarized by Vail,[35] who states "The precipitates formed by the reaction of the salts of heavy metals with alkaline silicates in dilute solution are not the result of the neat stoichiometric reactions describing the formation of crystalline silicates, but are the product of an interplay of forces which yield hydrous mixtures of varying composition and water content." These reaction products are usually noncrystalline and therefore very difficult to characterize structurally. They are most often described as hydrated metal ions associated with silica or silica gel. Iler[36] mentions "that many ions are held irreversibly on silica surfaces by forces still poorly understood in addition to ionic attraction." The composition and form of the metal "silicates" formed from metal ions and soluble silicates are functions of the conditions under which they are formed: temperature, concentration, addition rate, metal ion speciation, presence of other species, etc. In addition to the use of soluble silicates, metal silicates can be formed from the cementitious reactions in typical soluble silicate systems. Bishop[28] postulated that the observed low leaching of metals in cement-soluble silicate systems was due to the association of the metals with silica, possible as silicates. However, his work started with metals already in solution. In fact, a method of creating metal silicates from metal hydroxides is by deliberate dissolution of the metal species — hydroxide, carbonate, sulfide, etc. — in acid followed by reprecipitation by silicates or polymeric dissolved silica.[37]

In certain cases, metal carbonates are less soluble than their corresponding hydroxides. In cement chemistry, the natural formation of carbonates from carbon dioxide from the air is termed "carbonation". Cote[1] has shown that carbonate ion concentration in a system depends on both CO_2 partial pressure and pH. The carbonate species, CO_3^{2-} dominates at pH values larger than 10.3. The carbonation process at alkaline pH is

$$Me(OH)_2(solid) + H_2CO_3 \rightarrow MeCO_3(solid) + 2H_2O$$

The pH at which carbonation occurs depends on the solubility products of the carbonate and hydroxide species, and the CO_2 concentration. Patterson et al.[38] found that the formation of hydroxide precipitates controlled the solubility of zinc and nickel over a wide range of pH values, but cadmium and lead solubilities were controlled by carbonate precipitates in a narrower range. This is in keeping with the results of treatability tests conducted by the author on lead-bearing wastes, using soluble and "insoluble" carbonates. However, precipitation by addition of carbonates has not been widely used in either wastewater or S/S treatment. One problem is that the carbonates may be decomposed at low pH, such as that encountered in the TCLP test, if the pH of the leaching solution in contact with the solid actually drops too low. If CO_2 is evolved, the reaction is irreversible even if the final pH of the leachant is high, and the final speciation of the metal will be as the hydroxide. This may explain the widely varying results reported (informally) by S/S investigators testing carbonates, since the efficacy of carbonate precipitation would seem to be unusually sensitive to test variables; the sample being tested could either lose carbonate as CO_2 or gain it in the same way from the atmosphere (air) in the test container. Also, curing time may affect speciation and, therefore, leaching results in waste forms subject to atmospheric carbonation, especially over long time periods.

There appears to have been little work done on the use of phosphates for metal fixation, although ferric chloride is commonly used for phosphate removal in wastewater treatment. There seem to be possibilities here for those metals that form low-solubility phosphate species, but the general hazardous waste S/S literature contains no examples of the use of phosphates in these systems. Phosphate chemistry is very complex and varied. Compounds containing monomeric PO_4^{3-} are called orthophosphates or simply phosphates. The latter broad term is also used to describe all compounds in which phosphorus atoms are surrounded by a tetrahedron of four oxygen atoms. Since the oxygen atoms can also be shared among tetrahedra, chains, branched chains, and rings can be formed, in one-, two-, and three-dimensional networks. The linear formula can be expressed as $M_{n+2}P_nO_{3n+1}$. For orthophosphates, n = 1. Linear chains are known as polyphosphates for n = 2, 3,....(pyro- or dipoly-, tripoly-, etc.), cyclic rings as metaphosphates, and branched or cage structures as ultraphosphates. When n becomes very large, the composition becomes indistinguishable from that of the cyclic metaphosphates, $(MPO_3)_n$. The simple phosphate salts of the toxic metals have low water solubility, although they are soluble in acids. However, phosphates with n > 1 have the potential to sequester the metals as water-soluble species. This property is the basis for detergent and water treatment applications of phosphates. Therefore, the presence of phosphates in the waste or the S/S system may be harmful or beneficial to fixation, depending on the phosphate species.

One of the most intriguing prospects for metal immobilization is coprecipitation with other metal species. The removal of toxic metals from wastewater with systems which coprecipitate and/or flocculate them with iron and aluminum salts is well known and widely used.[39-41] More recently, it has been used to reduce the solubility of various toxic metals in hazardous waste treatment.[42] The ratio of Fe^{2+} to Fe^{3+} is important, with ratios of 1:1 to 1:2 reported as yielding optimum results. Sols of hydrous metal oxides are stabilized by the presence of excess ferric ion, but acquire a negative charge, destabilize, and flocculate under alkaline conditions. As the system becomes alkaline, the ferrous ion is also easily oxidized to ferric, and precipitates as the hydroxide. These reactions remove other metal ions from solution, reducing their concentrations to levels below those obtained with simple hydroxide precipitation. Other complexes can also produce insoluble metal species which may have lower solubilities than the simple metal compounds.

Many metals exist in ores and rocks as complex, crystalline structures with lower solubility than the noncrystalline structural isomers or hydrated homologues found in S/S systems. This is an area that needs exploration, because it may have important benefits in S/S work. Wastes and waste forms resulting from S/S often show changed leachability with aging, and this may very well be due to changes in molecular arrangement or morphological phenomena. For example, precipitation of metals under controlled temperature and pH conditions will favor the production of thermodynamically favored crystalline solids.[43] Such solids exhibit stronger chemical bonds and are thus more resistant to leaching. In other cases, the coprecipitation process can result in isomorphic substitution of metal atoms in another crystal structure, adsorption, and entrapment by occlusion or in pore spaces.[44] If mechanisms for the formation of low-solubility, complex metal species can be elucidated, it may be possible to create favorable conditions for them to operate. Also, since crystalline species can be studied by X-ray and electron diffraction techniques, their structures are much easier to determine than the amorphous compounds that we usually encounter in S/S.

Ordinarily, S/S investigators think in terms of fixation of metals with inorganic species. However, many organic compounds also form low-solubility species with certain metals. Manahan and Smith[45] report that humic acids formed in the decay of vegetable matter immobilize metal ions in sediments and soil. Saar and Weber[46] state that the "many oxygen-containing functional groups, particularly –COOH and –OH ...... enable fulvic acid to behave as a polyelectrolyte." Rather than precipitate metals from solution in the usual sense — by formation of low-solubility species from ions in solution — another approach is to react the metal with an active area or functional group on the surface of an insoluble substrate. Nelson[47] used a treated leather waste to remove heavy metal ions, particularly lead and cadmium, from nitrate and acetate solutions. Another process[48] uses modified casein to remove cadmium, chromium, copper, mercury, nickel, and zinc from wastewater. Chromium reportedly can be removed as directly as Cr^{6+}, without prior reduction, to levels below 1.0 ppm from streams containing up to 500 ppm Cr^{6+}. The casein is modified by treating with formaldehyde to form a cross-linked, insoluble product. The most widely publicized insoluble substrate for heavy metal fixation has been insoluble starch xanthate (ISX).[49,50] ISX is produced by treatment of starch with cross-linking agents, then xanthating it with CS_2 in the presence of an alkali metal base such as sodium hydroxide. The resulting product is a particulate solid with the structure

$$\left[Starch - O - \overset{\overset{\textstyle S}{\|}}{C} - S - \right]$$

In contact with metal ions, the metal links to the sulfur group much as it would with the S^{2-} in inorganic sulfides. Cellulose xanthates operate in much the same way.

The existence of metal species in a waste, or their precipitation during S/S treatment or during leaching of the waste form, is not always as large particles easily removed during laboratory filtration. Formation of colloids — i.e., very small, nonsettleable particles not in true solution — is a very common phenomenon in chemistry. It is also more common than previously believed in S/S systems, much to the chagrin of many investigators in the field. We have often observed the presence or formation of colloidal lead species which are not removed by the filter used in the TCLP (0.7 μ) but are removed to a much greater degree by the 0.45-μ filter used in the EPT. This is vividly illustrated by the following data from a study in which three filtration methods were used in a modified TCLP test:

Filter	Lead Concentration in Filtrate (mg/l)
Unfiltered	29.10
Regular TCLP filter	15.30
Whatman #42	0.04

The high-retention Whatman paper removes virtually all colloidal size particles, while the TCLP glass fiber filter does not. The Whatman paper was previously checked with true solutions of lead compounds to verify that it does not preferentially sorb lead from solution.

The question raised by these data is this: what filter is appropriate for use in leaching tests? The filters used in the TCLP and EPT were chosen for good retention and fast filtration, the latter property being improved by the change of filters from the EPT to the TCLP. At present, EPA's reaction to these facts is that any metal not removed by the filter is presumed to be mobile in the soil environment. However, this stance is apparently not supported by scientific data. Are colloids mobile in "soil," especially in view of the fact that soils are very variable materials in themselves? If so, at what size range, since the colloidal spectrum may be a continuum from true solubility to settleable particles in a given system? This phenomenon is probably the prime factor in the large differences sometimes seen between EPT and TCLP test results, mainly for lead. However, it may also operate for other metals and species.

Workers in the field have found methods to minimize the formation of colloidal lead by the use of additives or by aging of the waste form. However, the fundamental question of mobility raised by the presence of colloidal species has not been addressed. It may be very important when working with ground-water models where assumptions are made about the mobility of species in the geostructure. Also, perhaps more care needs to be exercised in specifying the filtration conditions used in leaching tests if they are intended to simulate any real set of conditions in the environment.

Adsorption   In a solid, molecules are held together by a variety of types and magnitudes of cohesive forces. Ionic, valence bonds are very strong. Covalent bonds are weaker, and van der Waals forces are weaker yet. At solid surfaces, cohesive forces are unbalanced, giving rise to the phenomenon of adsorption. Adsorption occurs when a molecule or ion becomes attached to a surface, usually as a monolayer. Therefore, the adsorbing ability of a material or system is directly related to its surface area, which in turn is a function of both particle size andshape, and porosity. The phenomenon is reversible in general, but the ease and completeness of desorption vary with the nature of the solid surface, and depend on factors such as concentration and pH. Adsorption is also highly specific as to species.

Adsorption is commonly described in terms of the Langmuir adsorption isotherm.[51] Many simple systems show Langmuir-type behavior; whether or not many S/S systems would do so, the concepts and mechanistic models resulting from adsorption theory are valuable in exploring actual and possible immobilization of constituents in S/S systems. Another adsorption isotherm by Freundlich is an empirical relationship which, in its linear form, is more often used in the analysis of experimental data. Adsorption is most effective at low concentrations of the species in solution where it is well described by the Langmuir and Freundlich models. Most research treats the phenomenon as one that reaches equilibrium rapidly. However, this is not necessarily true, especially with the desorption process. Therefore, the effectiveness of adsorption should be evaluated to take time into account.

Two major sorptive ingredients which can be used in S/S are activated carbon and clays. Adsorption on activated carbon is well documented in the literature.[52-54] It takes place through forces of predominantly physical nature — van der Waals forces — and desorption is relatively easy. Most use of activated carbon has been for adsorption of organics, although the material will also adsorb metal ions and other inorganics.[55]

The use of clays and other substrates for adsorption has been described by Chan et al.[56] and Theis et al.[57] Griffin et al.[58] found that adsorption of the cationic metals [lead, cadmium, zinc, copper, and chromium (III)] increased with increasing pH, while the opposite was true with anionic metals [chromium(VI), arsenic, and selenium]. Adsorption of cations by certain clays is caused by the negative surface charge which results from the clay structure itself. The resulting adsorption is not as reversible as that of simple surface adsorption,[59] and the mechanism is more that of chemisorption than of simple, Langmuir-type adsorption.

In surface layers of metal oxides, the metal ions have a reduced coordination number. In the presence of water, they develop a hydroxylated surface[60] with a net negative charge at high pH. This causes the adsorption of cations at any pH higher than that of zero point charge. The adsorption occurs before overall precipitation takes place, saturating the surface sites. This is a rapid, but reversible reaction. Since metal oxides are present in most S/S systems in the waste or in the matrix, their adsorption properties are of interest.

Chemisorption   Close-range chemical or physical forces can cause the retention of one species by another, usually at the surface of the latter. The binding forces are greater than those of pure adsorption, but do not represent a true chemical bond. There is no sharp boundary between adsorption and chemisorption, and some of the mechanisms described for adsorption really can be categorized as the latter. Functionally, the major difference is that chemisorbed species do not desorb easily.

Passivation   This phenomenon is important in S/S treatment. Metal ions dissolving from a solid surface may precipitate on the surface after contacting an anion in solution which forms a less soluble species. If the precipitate forms a tight, impermeable layer, it may block or inhibit further reaction at the surface. Such phenomena are highly specific to the constituents and the system as a whole, and may operate only temporarily. One such system is that of the formation of a silica gel on the surface of metals salts exposed to a soluble silicate solution. The gel layer prevents the passage of further metal ion into solution, thus slowing dissolution for a time. However, it also acts as a semipermeable membrane, allowing the passage of water which builds osmotic pressure on the solid side. Eventually, the membrane ruptures, spilling concentrated metal solution into the liquid phase. Depending on the relative concentrations of the constituents and their stoichiometry, and on pH, the metal may precipitate as a low-solubility silicate, as a hydroxide, or may remain in solution. In some cases, this process may repeat itself periodically, possibly giving rise to fluctuations in leaching rate with time.

Ion Exchange   A number of natural materials such as soils and zeolites, as well as various synthetics, have the ability to exchange ions which they contain for others in solution. The most important are cation exchangers and their ability to do this is expressed as cation exchange capacity (CEC), the magnitude of which determined by the charge of the hydrated cation and its size in solution. Removal efficiency of an ion from solution is directly proportional to its charge and inversely proportional to its size. Organic synthetic ion exchangers contain acidic groups such as -COOH, -SO$_3$H, and -OH. Natural and synthetic inorganic ion exchangers may exchange H^+ or a monovalent cation such as Na^+ for polyvalent metals in solution.

The charge on a clay surface may arise from either of two sources:[32] (1) ionization of the hydroxyl groups attached to silicon atoms at the surface and (2) isomorphous substitution. The latter phenomenon is the result of substitution of Al^{3+} and Fe^{3+} for Si^{4-} or Mg^{2+} for Al^{3+}. Isomorphous substitution results in a fairly uniform charge distribution over the surface of the clay particle in the "2:1" class clays such as montmorillonite, beidellite, and vermiculite. 2:1 clays are composed of layers consisting of one alumina sheet between two sheets of silica. The other class of clays, "1:1" (one layer each of silica and alumina), has a lower surface charge resulting from hydroxyl group ionization.

The deliberate addition of ion exchange materials to S/S formulations has had limited use, but common reagents often have limited ion exchange properties: clays and other soils, cements, and flyashes. Synthetic resins are usually too expensive for most S/S use, primarily because the resin also will exchange ions in solution other than the toxic metals, e.g., calcium. This has another negative effect: it may inhibit solidification reactions such as cement setting byremoving calcium before it can take part in the setting reactions (see Chapter 10). Anion exchange also can take place. This type of reaction is favored by low pH and is much greater in 1:1-type clays.

Diodochy   When one element substitutes for another of similar size and charge in a crystalline lattice, it is called diodochy.[61] Examples are zinc or manganese for calcium in calcite, or arsenic for phosphorus in apatite. Bishop[27] postulates that chromium and lead are bound into the silica matrix in Portland cement waste forms, possibly as "silicates". Whether these elements actually replace silicon in the lattice is questionable, but it appears that they coexist with silica, perhaps as hydrated metal oxides, at the molecular level rather than merely being microencapsulated. The latter mechanism is in keeping with theories about the actual structure of heavy metal silicates.[36,37] It is similar to diodochy except that the substitution is by metal oxides embedded in the silica lattice rather than ion for ion replacement. Coprecipitation, as with iron in water treatment,[62] might also be very similar.

Reprecipitation   This may be an important containment mechanism for metals in sludges, where they are usually originally speciated as hydroxides and then subjected to acid leaching. It is generally assumed that low-solubility species such as hydroxides do not respeciate. If this were the case, they should leach at rates approximated by the respective hydroxide solubilities at any given pH, especiallyunder the conditions prevalent in a leaching test such as the TCLP. Even allowing for other fixation mechanisms such as those presented previously, the shape of the leaching vs. pH curve should be the same.

In fact, this is not the case in cement- and pozzolan- based systems. One explanation is that, as the acid leachant dissolves alkali from the solid matrix and thus creates a localized acidic condition, metal hydroxide dissolves. Subsequently, the metal is reprecipitated by one of the mechanisms which we have just discussed. Another possibility is that the metal is reprecipitated before the leaching takes place during pozzolanic reactions or in the formation of cement paste in cement-based systems. In both systems, similar reaction steps occur, the first of which is the formation of highly alkaline conditions as calcium dissolves from the cement or lime. This results in the dissolution of certain of the metal hydroxides at the same time that silica is being solubilized as the silicate. The heavy metals then compete with the alkali earth metals such as calcium for the silicate, both precipitating the heavy metals and interfering with the normal setting reaction. The latter effect is known in cement chemistry.[63]

Encapsulation   In the early days of S/S development, the word "encapsulation" was used to describe many of the processes. If encapsulation is the only operative mechanism, then the degree of protection to the environment is determined entirely by the matrix and its continuing ability to isolate the waste from the environment. Isolation, in turn, depends primarily on the permeability of the matrix. Encapsulation without any other fixation mechanism rarely occurs in commercial S/S operations.

Alteration of Waste Properties   Frequently, the highest degree of containment can be obtained by pretreating the waste in some manner, altering its properties before S/S. Examples include reduction of hexavalent chromium to the less soluble trivalent state, destruction of cyanide, and breaking of soluble nickel complexes. In some cases, particle size reduction to allow better reaction of the waste with the reagents may be useful. In others, dewatering may provide a better final waste form at lower cost. Often, removal of immiscible organic phases is necessary to achieve acceptable leaching and meet recent land bans based on total constituent analysis. Acids are usually neutralized, at least partially, prior to S/S. In some cases, it may be necessary to lower the pH of caustic wastes to meet RCRA corrosivity requirements.

Leaching Mechanisms

Leaching mechanisms which control and are controlled by these containment systems are

- Solubilization
- Transport through the solid
    Convective transport
    Molecular diffusion
        Solid state
        Pore solution
- Transport through the solid-liquid boundary
- Bulk diffusion in the liquid leachant

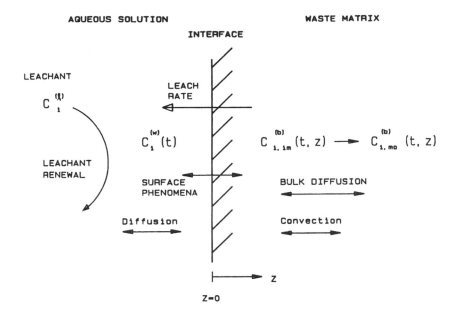

Figure 2-8.   Leaching mechanisms. (From Cote, P. L. Personal commu-
               nication.)

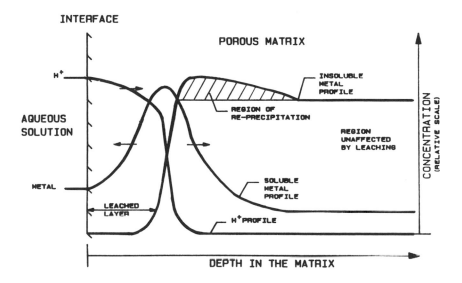

Figure 2-9.   Concentration gradients during leaching. (From Cote, P.
               1986. "Containment Leaching from Cement-Based Waste
               Forms under Acidic Conditions." Ph.D. thesis, McMaster
               University, Toronto.)

• Chemical reactions in the leachant
• Biological attack

These mechanisms are illustrated in Figure 2-8 schematically.[64] Figure 2-9 presents the solid-state mechanisms in another way by plotting the various concentration gradients vs. position in the solid matrix. For most S/S matrices in contact with neutral groundwater, the leaching rate will be controlled by molecular diffusion of the solubilized species. Under acidic conditions, the rate is initially limited by the supply of H^+. After a time, molecular or boundary layer diffusion of the waste constituents again becomes rate limiting since H^+ diffuses much faster than any other species.[16]

Even in the case of a finely divided solid and a rapidly agitated leaching solution as, for example, in the EPT or TCLP test methodologies, the leachant requires 18 to 24 h to approach equilibrium constituent concentrations for metals. Under real leaching conditions, equilibrium is attained much more slowly. If the leachant flows slowly around or through the waste form, near equilibrium conditions may develop. Under faster flow conditions, transport of constituents may be solubility limited. In the former case, molecular diffusion through the porewater will probably be the limiting factor, since it is likely to be more rapid than true solid-state diffusion. This scenario is the more likely one in a properly managed hazardous waste landfill, e.g., one meeting RCRA Minimum Technology Requirements (MTR). This is true, however, only after the landfill cell is closed so that water inflow is minimal. When the cell is "open", i.e., being filled, excessive rainfall can create rapid flow conditions through the waste, and the leachate will not become saturated.

Solubilization   For any constituent to leach, it must first dissolve in the pore water of the solid matrix or in the leachant permeating the solid. Some species dissolve more slowly than others, with the rate of solubilization being controlled both by basic solubility considerations previously discussed and by the concentration in the solution near the surface. The latter, in turn, is controlled by boundary layer transport phenomena. The thinner the boundary layer, the faster the rate of dissolution, until the bulk leachant is saturated.

Transport through the Solid   Within the solid waste form, transport could occur either by advection or by diffusion. Most waste forms have relatively low permeability, so advective transport is probably not important in either a leaching test or in actual field conditions. Therefore, diffusion is the only effective process operating.

Molecular diffusion — Diffusion occurs by the random motion of individual molecules or ions. Assuming that the solid is in chemical equilibrium when leaching begins, diffusion is driven by the difference in chemical potential (constituent concentration) between the solid and the fluid leachant. The chemical gradient thus created causes constituents to migrate from the solid to the leachant. The flux of the constituent at a position within the solid is described by Fick's first law.

$$J = -D \frac{dC}{dz} \qquad (5)$$

where C = concentration (mass/unit volume of solution)
        D = diffusion coefficient (area of solution/time)
        J = flux (mass/area of solution $\times$ time)
        z = distance (L)

The "area of solution" for a porous solid is not known, so the flux is normally expressed on an "area of porous solid" basis. Cote[1] has treated this more rigorously, including a term for connected porosity, and also a modification for tortuosity. The later term expresses the fact that diffusion in a porous solid "takes place in a tortuous path of fluid between and around solid particles." A widely accepted model for the leaching of S/S treated wastes was proposed by Godbee et al.[65] Godbee's model can be combined with the formula used in the American Nuclear Society's (ANS) leaching protocol,[66] which is a series of seven sequential batch leachings. This gives a model that accounts for most of the variables which effect metal leaching from S/S treated waste. These include constituent speciation, particle size, and the initial concentration of the constituent. However, boundary layer effects and chemical reactions occurring during the leaching process can significantly effect the practical utility of such models.

These mathematical models can be used to determine leaching rates which, when coupled with data about the disposal site and groundwater, can be used to predict *in situ* leaching rates over extended time periods. A number of groundwater models have been developed and used for such predictions, including the EPA VHS (vertical and horizontal spread) model. All mathematical models, however, should be used with caution because of the simplifications utilized in their development, and because of other factors not considered nor even recognized at the present stage of development.

As more soluble constituents are leached from a relatively insoluble solid matrix, a layer deficient in the leached constituents develops.[12,16] Under low pH conditions, both H^+ and the leachable constituents must diffuse through this layer in opposite directions. As we saw earlier, the leaching rate in the leached layer should eventually be limited by diffusion of constituents, since H^+ diffuses muchfaster than other species. However, this layer may not be rate limiting in the overall leaching process. As constituents leach, the layer may become more porous compared to the unleached solid, so that molecular diffusion in the pore water, and boundary layer phenomena, become the limiting factors.

Transport through the Solid-Liquid Boundary   In many leaching scenarios, the solid-liquid interface may create the limiting condition for leaching, as it does in electrochemical phenomena. The speed with which a system comes to equilibrium is frequently limited by the thickness and nature of the boundary layer. In attempting to develop fast leaching assessment tests for routine quality control, for example, this limitation is thought to be controlling.[67] This layer is frequently described as the "electrical double layer". At any interface between two phases, there is an unequal distribution of electrical charges, with a net negative charge on one side and a corresponding net positive charge on the other. Because overall electrical neutrality must be maintained, the net charges must equal each other. This localized unequal distribution gives rise to a potential across the interface. The phenomenon was first treated mathematically by Von Helmholtz,[68] later modified to the Gouy-Chapman model, and finally by Stern,[69] whose model is still used today.

Stern divided the solution side of the double layer into two sublayers: (1) an inner region — now called the "Stern layer" — of strongly adsorbed counterions about one hydrated ion radius thick at fixed sites, corresponding to Langmuir-type adsorption and (2) a diffuse region of counterions distributed according to electrical forces and thermal motion. The electrical potential drops rapidly through the Stern layer, possibly even changing the sign resulting from the original charged surface, and then more gradually through the diffuse layer. The shear plane between the surface and the surrounding solution is somewhere in the diffuse layer, and the potential difference between the shear plane and the solution is known as the zeta, or electrokinetic, potential. Zeta potential is a measurable value widely used in colloid and surface chemistry.

The effective thickness of the diffuse portion of the double layer is called the Debye length. It is important to our discussion of leaching because any species moving from the solid into solution must traverse this layer. The Debye length is inversely proportional to the valence of ions in the solution, and directly proportional to the square roots of the absolute temperature and dielectric constant. In water, with a high dielectric strength, electrical effects extend relatively far into solution; in the presence of an electrolyte, the layer is compressed.

Bulk Transport in the Liquid Leachant   The effects of bulk transport in the leachant — or in the environment, the groundwater — depend primarily on the hydraulic regime. In batch tests, the system approaches equilibrium and the leaching rate decreases as the chemical potential between the solid and the solution phases — the driving force — decreases to near zero. This may also occur in a landfill which is saturated, but with no movement of groundwater. It results in the highest possible concentration of constituents in the leachate, but the overall leaching rate may be lower than in a dynamic system. Fast-moving groundwater (or a column leaching test in the laboratory), on the other hand, will transfer constituents from the solid as rapidly as the solid and boundary layer transport mechanisms will allow. With a highly soluble constituent in a porous matrix, leaching rate can be very rapid.

Chemical Reactions   The various scenarios and mechanisms which we have discussed to this point assume that no chemical reactions (other than those involved in dissolution of the constituents in the solid) occur. While this may be true in laboratory leaching tests, since they were so designed, it is not the case in the real environment. Rain, surface, and groundwaters all contain constituents which may increase or decrease the leaching rate. Anions such as carbonate, sulfide, and silicate, organic chelating agents, and adsorptive particulates all can affect leaching. Precipitates may passivate waste form particles, slowing or completely blocking transport through the solid or the boundary layer by clogging up the pores. On the other hand, chelating agents can increase solubility by preventing precipitation and converting ionic species into soluble organic complexes, preventing saturation in the leachant of the ionic species. Oxidizing conditions can increase the solubilization rate of some species, while a reducing environment may do the same for others.

Biological Attack   In the long term, potential biological attack of S/S waste forms is a matter of concern environmentally. Most S/S systems produce conditions both of high pH and high total alkalinity. These conditions are not conducive to the activity or even survival of most microorganisms, as demonstrated by the long-standing practice of "stabilizing" sewage sludges by addition of lime. Organic S/S systems such as urea-formaldehyde (UF), on the other hand, have been shown to be biodegradable. For most waste forms, the primary concern has been that of indirect attack through the known biological mechanisms which produce acids that in turn can attack certain waste constituents or even the matrix itself. Plant roots produce carbonic acid when alive, and organic acids when decaying. Plant roots, however, are not normally in contact with hazardous wastes, treated or otherwise, in properly managed disposal scenarios.

A more reasonable concern is that of the biologically enhanced[70] oxidation of insoluble metal sulfides to soluble sulfates, releasing the metal and also generating sulfuric acid which can attack the matrix. This reaction is partly responsible for the acid mine drainage from iron sulfides. It requires, however, oxidizing conditions that are not normally present in hazardous waste landfills after completion of the cover system. In uncontrolled sites, these conditions could occur near the surface, especially with low-alkalinity systems.

2.5. SUMMARY

This discussion of the chemistry of S/S waste formsconsidered three basic elements: the waste, the S/S reagents, and the environment in which the waste form will exist. These elements were discussed in terms of three chemical interaction subsets:

*   The system as an adulterated, weak "concrete," with the waste providing both water (in most cases) and "aggregate".
*   The system as a treated waste, with the S/S reagents being merely treatment chemicals in the same sense as in the wastewater field.

•   The system in which the waste-S/S reagent combination interacts with its disposal or end-use environment.

After achieving some understanding of the first two approaches, the latter system was explored by first discussing the measurement of leachability and the factors which affect it. The immobilization/containment mechanisms were then examined in detail, followed by discussions of the inverse situation — the leaching mechanisms. Several illustrations of containment mechanisms and their effects in real systems were presented using experimental data generated in the author's laboratory. These concentrated on solubility of metals and its relationship with pH.

REFERENCES

1.   Cote, P. 1986. "Contaminant Leaching from Cement-Based Waste Forms under Acidic Conditions." Ph.D. thesis, McMaster University, Toronto.

2.   Conner, J. R., 1990. *Chemical Fixation and Solidification of Hazardous Wastes.* Van Nostrand Reinhold, New York. pp. 293–298.

3.   Double, D. D. and A. Hellawell. 1977. "The Solidification of Cement." Sci. Am. 237: 82–90.

4.   Conner, J. R. 1990. *Chemical Fixation and Solidification of Hazardous Wastes.* Van Nostrand Reinhold, New York. pp. 343-346.

5.   Double, D. D., W. L. Thomas, and D. A. Jameson. 1980. "The Hydration of Portland Cement; Evidence for an Osmotic Mechanisms." *Proc. 7th Int. Congress on Chem. Cement.*

6.   Skalny, J. and K. E. Daugherty. 1972. "Everything You Always Wanted to Know About Portland Cement." *Chemtech* 38–45.

7.   Portland Cement Association, Skokie, IL 1974. "Concrete Information."

8.   Kirk-Othmer. 1979. *Encyclopedia of Chemical Technology,* 3rd ed. John Wiley & Sons, New York.

9.   Popovics, S. 1970. *Concrete Making Materials.* McGraw-Hill, New York.

10.   Lea, F. M. 1970. *The Chemistry of Cement and Concrete.* Edward Arnold, London.

11.   Cullinane, M. J., R. M. Bricka, and N. R. Francingues. 1987. "An Assessment of Materials that Interfere with Stabilization/Solidification Processes" in Proc. 13th Annual Research Symposium. U.S. Environmental Protection Agency, Cincinnati. pp. 64–71.

12.   U.S. EPA. 1988. Best Demonstrated Available Technology (BDAT) Background Document for F0065, Electroplating." U.S. Environmental Protection Agency, Washington, DC.

13. Ashworth, R. 1965. "Some Investigations into the Use of Sugar as an Admixture to Concrete." *Proc. Inst. of Civil Engineering.*

14. Chalasani, D., F. K. Cartledge, H. C. Eaton, M. E. Tittlebaum, and M. B. Walsh. 1986. "The Effects of Ethylene Glycol on a Cement-Based Solidification Process." *Haz. Wastes Haz. Materials* 3(2).

15. Roberts, B. K. 1973. "The Effect of Volatile Organics on Strength Development in Lime Stabilized Fly Ash Compositions." M.Sc. thesis, University of Pennsylvania, Philadelphia.

16. Rosskopg, P. A., F. J. Linton, and R. B. Peppler. 1975. "Effect of Various Accelerating Chemical Admixtures on Setting and Strength Development of Concrete." *J. Testing and Evaluation* 3(4).

17. Smith, R. L. 1979. "The Effect of Organic Compounds on Pozzolanic Reactions." I.U. Conversion Systems Report No. 57, Horsham, PA.

18. Walsh, M. B., H. C. Eaton, M. E. Tittlebaum, F. K. Cartledge, and D. Chalasani. 1986. "The Effect of Two Organic Compounds on a Portland Cement-Based Stabilization Matrix." *Haz. Wastes Haz. Materials* 3(1).

19. Young, J. F. 1972. "A Review of the Mechanisms of Set-Retardation of Cement Pastes Containing Organic Admixtures." *Cement Concrete Res.* 2(4).

20. Young, J. F., R. L. Berger, and F. V. Lawrence. 1973. "Studies on the Hydration of Tricalcium Silicate Pastes. III. Influences of Admixtures on Hydration and Strength Development." Cement Concrete Res. 3(6).

21. 1974. "Safety of Public Water Systems (Safe Drinking Water Act)." Public Law 93-523.

22. Darcel, F. 1984. "Recent Studies on Leach Testing — A Review." Ministry of Environment, Ontario.

23. U.S. EPA. 1985. "Solid Waste Leaching Procedure Manual." U.S. Environmental Protection Agency, Washington, DC. (EPA SW-924).

24. Cote, P. L. and D. Isabel. 1984. *Hazardous and Industrial Waste Management and Testing: Third Symposium.* American Society for Testing and Materials, Philadelphia. pp. 48–60.

25. "Guide to the Disposal of Chemically Stabilized and Solidified Waste." U.S. Environmental Protection Agency, Washington, DC. (EPA SW-872).

26. Conner, J. R. 1990. *Chemical Fixation and Solidification of Hazardous Wastes.* Van Nostrand Reinhold, New York. pp. 26–32.

27. Conner, J. R. 1990. *Chemical Fixation and Solidification of Hazardous Wastes.* Van Nostrand Reinhold, New York. pp. 63–71.

28. Bishop, P. L. 1988. "Leaching of Inorganic Hazardous Constituents from Stabilized/Solidified Hazardous Wastes." *Haz. Waste Haz. Materials* (5)2: 129–143.

29. Cote, P. 1986. "Contaminant Leaching from Cement-Based Waste Forms under Acidic Conditions." Ph.D. thesis, McMaster University, Toronto.

30.   1987. Federal Register Vol. 52, No. 155: 29999.
31.   Moore, J. N., W. H. Ficklin, and C. Johns. 1988. "Partitioning of Arsenic and Metals in Reducing Sulfidic Sediments." *Environ. Sci. Technol.* 22: 432–437.
32.   Dragun, J. 1988. "The Fate of Hazardous Materials in Soil," *Haz. Materials Control* 41–65.
33.   Sandesara, M. D. 1978. U.S. Patent 4,118,243.
34.   Conner, J. R. 1974. "Ultimate Disposal of Liquid Wastes by Chemical Fixation." Proc. 29th Annual Purdue Industrial Waste Conference, Purdue University.
35.   Vail, J. G. 1952. *Soluble Silicates.* Van Nostrand Reinhold, New York
36.   Iler, R. K. 1979. *The Chemistry of Silica.* John Wiley & Sons, New York.
37.   Rousseaux, J. M. and A. B. Craig, Jr. 1980. *Stabilization of Heavy Metal Wastes by the Soliroc Process.* Cemstobel, Brussels.
38.   Patterson, J. W., H. E. Allen, and J. J. Scala. 1977. "Carbonate Precipitation from Heavy Metals Pollutants." *J. Water Pollution Control Federation* 12: 2397–2410.
39.   Sittig, M. 1973. *Pollutant Removal Handbook.* Noyes Data Corporation, London.
40.   LeGendre, G. R. and D. D. Runnells. 1975. "Removal of Dissolved Molybdenum from Wastewaters by Precipitates of Ferric Iron." *Environ. Sci. Tech.* 9(8): 744–749.
41.   Swallow, K. C., D. N. Hume, and F. M. M. Morel. 1980. "Sorption of Copper and Lead by Hydrous Ferric Oxide." *Environ. Sci. Tech.* 14(11): 1326–1331.
42.   Pojasek, R. B. 1980. *Toxic and Hazardous Waste Disposal,* Vol. 1–4. Ann Arbor Science Publishers, Ann Arbor, MI.
43.   Feitknecht, W. and P. Schindler. 1963. "Solubility Constants of Metal Oxides, Metal Hydroxides and Metal Hydroxide Salts in Aqueous Solution," *Pure Applied Chem.* 6: 132–192.
44.   Leckie, J. O., M. M. Benjamin, K. Hayes, G. Kaufman, and S. Altman. 1980. Adsorption/Coprecipitation of Trace Elements from Water with Iron Oxyhydroxide. (EPRI CS-1513), Stanford University, CA.
45.   Manahan, S. E. and M. J. Smith. 1973. "The Importance of Chelating Agents." *Water and Sewage Works* 102–106.
46.   Saar, R. A. and J. H. Weber. 1982 "Fulvic Acid: Modifier of Metal-Ion Chemistry." *Environ. Sci. Tech.* 16(9): 510a–517a.
47.   Nelson, D. A. 1980. "Removal of Heavy Metal Ions from Aqueous Solution with Treated Leather." American Chemical Society, Houston.
48.   1979. *Chem. Eng.* 83–84.
49.   Wing, R. E. 1975. "Corn Starch Compound Recovers Metals from Water." *Industrial Wastes* 26–27.
50.   Wing, R. E. 1977. U.S. Patent 3,294,680.

51. Rosen, M. J. 1978. *Surfactants and Interfacial Phenomena*. John Wiley & Sons, New York.

52. Chereminisoff, P. N. and F. Ellerbusch. 1980. *Carbon Adsorption Handbook*. Ann Arbor Science Publishers, Ann Arbor, MI.

53. Perrich, J. R., Ed. 1981. *Activated Carbon Adsorption for Wastewater Treatment*. CRC Press, Boca Raton, FL.

54. Suffet, I. H. and M. J. McGuire. 1981. *Activated Carbon Adsorption of Organics from the Aqueous Phase*. Ann Arbor Science Publishers, Ann Arbor, MI.

55. Huang, C. P., P. K. Wirth, and D. W. Blankenship. 1981. "Removal of Cd(II) and Hg(II) by Activated Carbon." Proc. 2nd Natl. Conf. on Environmental Engineering. American Society of Civil Engineers, Atlanta. pp. 382–390.

56. Chan, P. C., J. W. Liskowitz, A. Perna, and P. Trattner. 1980. "Evaluation of Sorbents for Industrial Sludge Leachate Treatment." U.S. Environmental Protection Agency, Cincinnati.

57. Theis, T. L., J. L. Wirth, R. O. Richter, and J. J. Marley. 1976. "Sorptive Characteristics of Heavy Metals in Fly-Ash-Soil Environments." *Proc. 31st Industrial Waste Conference* Purdue University.

58. Griffin, R. A., R. R. Frost, and N. F. Shimp. "Effect of pH on Removal of Heavy Metals from Leachates by Clay Minerals." Illinois State Geological Survey, Urbana, IL.

59. Kinniburgh, D. G. and M. L. Jackson. Illinois State Geological Survey, Urbana, IL.

60. Schindler, P. W. 1981. *Adsorption of Inorganics at a Solid-Liquid Interface*. Ann Arbor Science Publishers, Ann Arbor, MI.

61. "Immobilization and Leachability of Hazardous Wastes." Environ. Sci. Technol. 16(4): 219a–223a.

62. Grutsch, J. F. 1983. *The Chemistry and Chemicals of Coagulation and Flocculation*. (American Petroleum Institute Manual on Disposal of Refinery Wastes).

63. Cullinane, M. J., R. M. Bricka, and N. R. Francingues, Jr. 1987. "An Assessment of Materials that Interfere with Stabilization/Solidification Processes." *Proc. 13th Annual Research Symposium*. Cincinnati. pp. 64–71.

64. Cote, P. L., T. R. Bridle, and A. Benedek. 1985. "An Approach for Evaluating Long Term Leachability from Measurement of Intrinsic Waste Properties." *3rd Int. Symp. Industrial and Hazardous Waste*. Alexandria, Egypt.

65. Godbee, H. et al. 1980. "Application of Mass Transport Theory to the Leaching of Radionuclides from Solid Waste." *Nuclear Chem. Waste Management* 1:29.

66. Measurement of the Leachability of Solidified Low-Level Radioactive Wastes. American Nuclear Society.

67.   Cote, P. L. Personal communication.
68.   Von Hemholtz, H. 1879. *Wied. Ann. Phys.* 7: 337.
69.   Stern, O. 1924. *Z. Electrochem.* 30: 508.
70.   1971. "Inorganic Sulfur Oxidation by Iron Oxidizing Bacteria." U.S. Government Printing Office, Washington, DC.

3 THE CHEMISTRY OF CEMENTITIOUS SYSTEMS FOR WASTE MANAGEMENT: THE PENN STATE EXPERIENCE

D. M. Roy and B. E. Scheetz

1.1. ABSTRACT

The Cement and Concrete Research Group at the Materials Research Laboratory of The Pennsylvania State University has had a long involvement with the national programs in nuclear waste management. This chapter highlights the results of much of this experience also in the context of other studies. Illustrations are given of the design of cementitious systems for waste management, emphasizing control of the chemical composition, particle characteristics, and other important features of the initial system. Relationships between the initial composition and design and important characteristics of the consolidated waste form such as porosity, pore size distribution, pore fluid chemistry, phase stability, permeability, and ionic transport are discussed. The importance of designing a system having both low permeability and containing relatively insoluble low leachability phases is emphasized, which together operate to restrict the transport of ions and species into and out of the waste system. Various forces of degradation of cementitious wasteforms are discussed, including the transport of harmful species (ions, fluids) into the cementitious material resulting in degradation of the structure. The counterpart, leaching of major constituents such as portlandite, gives rise to a weakening of the cement paste structure and, at the same time, increases the pore size distribution, providing a pathway for increased transport of harmful species out of the waste form, decreasing its effectiveness as a waste encapsulant.

The unique contribution from our research has been the developments leading to the fundamental control of cement chemistry which in turn governs the behavior of most other desirable properties: density, permeability, porosity, durability and both the mechanical and physical behavior of the composite materials.

1.2. INTRODUCTION

The durability of cementitious materials in waste disposal applications (encapsulants, sealants) is most directly affected by the presence and transport of fluids to and from the cementitious matrices and not, as is widely held, related to the strength of the material, although intrinsically strong cementitious bodies tend to possess low porosity and hence limit the access of degrading fluids and the species they carry into and from the interior of the bodies. The design of a durable cementitious material in the broadest sense consists of two aspects: control over the physical behavior of the material and control over the chemical behavior of the hydrating matrices. In practical materials design, neither can be totally isolated from the other. Control of the physical properties of cementitious materials involves a careful control of particulate size distributions of the system's components, placement, and processing variables. In contrast, control over the hydration behavior involves careful selection of components based upon bulk chemistry and chemical reactivity, as well as the wasteforms' interaction in the environment where it will be used. Synergistic effects are common and can often be used to enhance the performance of the material.

In order to enhance durability, fluids and ions must be prevented or inhibited from entering or leaving a cementitious body. Low permeability and leach resistance are major rate controlling properties of the matrix. Low permeability can be associated with either or both low porosity or more importantly a pore size distribution fine enough to minimize the transport of fluids and ions. This implies that cementitious matrices can have relatively large porosities and can still be "impermeable". Further, low permeability can be controlled by the distribution of sizes of the constituent components of the cement, the nature of the hydraulic reactivity (fast vs. slow), the interactions with other components of the cementitious system and processing as well as the use of processing aids. Insolubility of the matrix combined with low permeability prevents or retards the release of hazardous nuclides to the leachate. Such species are normally transported through the pore fluid under the influence of a concentration gradient. Relative insolubility of the cementitious matrix is related to the chemical composition and also mechanically to exposed surface area. Processing controls the surface area, while bulk chemical composition and chemical reactivity of constituents control the reaction products and the chemical stability.

1.3. DEGRADATION

External attack upon a cementitious wasteform primarily takes the form of movement of cations/anions into or out of the object. Direct alteration of the cementitious matrix can occur via solid-state diffusion, which for the most part is an extremely slow process. More realistically, ingress of aggressive ions occurs through the pore structure under the influence of a concentration gradient, in much the same manner as ions are released from the wasteform to the environment. Degradation of the matrix is accomplished by the crystallization of phases in the interior of the wasteform which exert pressure upon the matrix in excess of its tensile strength resulting in cracking. Reactions may occur with the major cement paste matrix components calcium-silicate-hydrate (C-S-H) or with other components, particularly calcium hydroxide or calcium aluminate hydrates. Chemical reactions which result in degradation, primarily from the formation of four compounds (ettringite, brucite, gypsum, and calcite) are most commonly responsible. These occur primarily through the reaction with various ions and species transported from the environment into the pore fluid. Degradation may also occur by means of transport of species out of the cementitious matrix, i.e., leaching.

Attack in Aqueous Media

Sulfate

Sulfate attack can be initiated from a number of sources; either sodium sulfate, calcium sulfate, or magnesium sulfate. The latter is a more aggressive form.

Calcium   Calcium sulfate in solution interacts with the tricalcium aluminate of the anhydrous cement or from a hydrous calcium aluminate hydration product to form calcium sulfoaluminates. The calcium sulfate reaction may also derive secondarily from the reactions of other sulfates such as Na_2SO_4 or $MgSO_4$ with $Ca(OH)_2$ from the cement paste. This reaction may take place in a stepwise reaction, first to form tetracalcium aluminate monosulfate 12-hydrate and upon further exposure to calcium sulfate, to form ettringite. These reactions are summarized:

$$3CaO \cdot Al_2O_3 \cdot 6H_2O + CaSO_4 + 6H_2O \rightarrow 3CaO \cdot Al_2O_3 \cdot CaSO_4 \cdot 12H_2O \quad (1)$$

$$4CaO \cdot Al_2O_3 \cdot xH_2O + 3CaSO_4 \cdot 2H_2O + H_2O \rightarrow 3CaO \cdot Al_2O_3 \cdot$$
$$3CaSO_4 \cdot 32H_2O + Ca(OH)_2 \quad (2a)$$

$$3CaO \cdot Al_2O_3 \cdot CaSO_4 \cdot 12H_2O + 2CaSO_4 + 20H_2O \rightarrow$$
$$3CaO \cdot Al_2O_3 \cdot CaSO_4 \cdot 32H_2O \qquad\qquad (2b)$$

Note that in these reactions calcium sulfate is an essential component.
Magnesium   Similar reactions will of course account for the formation of
ettringite from the sulfate supplied by the magnesium salt. Additionally, however, magnesium ions in solution may either react with calcium hydroxide to
form additional calcium sulfate as above or will attack the C-S-H gel of the
matrix of the wasteform and release calcium ions into solution which are
essential to react with the sulfate ion forming ettringite. Additionally the
magnesium ion reacts with hydroxyl to form insoluble magnesium hydroxide,
brucite. These reactions are summarized:

$$C - S - H + Mg^{2+} \rightarrow Ca^{2+} + Mg(OH)_2 + SiO_2$$

Brucite possesses a molar volume which is 100% greater than periclase, MgO,
resulting in primary breakdown of the matrix and substantial additional internal
stresses.

   Atkinson and Hearne[1] and Walton et al.[2] have reported models of the same
form for the attack of magnesium sulfate:

$$X = 0.34[C_3A / 8][(Mg) = (SO_4) / 0.19]t$$

where   X = depth of penetration
         C_3A = weight % of tricalcium aluminate in cement
         (Mg) = molar concentration of magnesium ion
         (SO_4) = molar concentration of sulfate ion
         t = time in years.

Sodium   Though not generally considered as destructive as magnesium, sodium sulfate will react with portlandite to form additional gypsum which is
then available for further reactions to form ettringite. However, the pH remains
high in contrast to the Mg situation:

$$Ca(OH)_2 + Na_2SO_4 + 2H_2O \rightarrow CaSO_4 \cdot 2H_2O + 2NaOH$$

Surface spalling or scaling may be observed.

Chloride

   Freidel's salt, $3CaO \cdot Al_2O_3 \cdot CaCl_2 \cdot 10H_2O$, possesses an open structure similar to that of $3CaO \cdot Al_2O_3 \cdot Ca(OH)_2 \cdot 12H_2O$ or monosulfate and may cause
internal stresses when formed within a hardened cementitious body by a

percolating solution.[3] The effects of the formation of Freidel's salt are not as pronounced as those of ettringite[4] and it is stable to temperatures in excess of 80°C. The usual concern associated with chloride ion presence is of course its effect on corrosion of any reinforcing steel.

Leaching of Portlandite

Portlandite, $Ca(OH)_2$, is a primary hydration product of both alite (Ca_3SiO_5) and belite (Ca_2SiO_4). The solubility of portlandite is retrograde in character and the solubility is the highest of all of the calcium containing phases in the cementitious system. Lea[5] suggests that as much as 25% of the calcium in hydrated Portland cement is in this form. The progressive removal of this relatively soluble phase will further enhance the accessibility of the open interconnected porosity to the surface which will accelerate leaching of the hydrated calcium silicate matrix and decrease the length of the diffusion pathways for the contents of the pore fluids. Furthermore, it will also affect the strength of the cementitious body. Lea reports a loss of 1.5% in strength for each 1% of the calcium removed from the matrix. Another accompanying effect when exposed to natural groundwaters is commonly carbonation,[6] although this does not always result in increased permeability.[7]

Alkali-Aggregate Reaction

Alkali-aggregate reactions may take place on imbibition of water, but may commonly be avoided by the selective use of nonexpanding aggregates and hence will not be dealt with here.[7,8]

1.4. PORE SYSTEM

Scheetz et al.[9] and Shi et al.[10] have demonstrated that the size distribution of pores in cements and concretes can be modeled. However, they have both noted that the modeling can not be accomplished by a single function. Scheetz et al.[9] demonstrated that using two functions, a log-normal and an exponential power function, typical mercury intrusion porosimetry (MIP) data can be adequately modeled. Further they showed that the crossover point for these two modeling functions occurred in the region usually defined as the upper limits of capillary porosity and that the crossover position varied as a function of the curing conditions of the samples. The authors described the larger porosity as extrinsic porosity introduced into the cementitous sample by mixing and handling of the pastes, while the smaller porosity was described as intrinsic porosity resulting from hydration processes. Shi et al.[10] have modeled similar data but have used three log-normal distributions to fit the MIP data.

The "macro" porosity is equivalent to the extrinsic porosity introduced by mixing and handling. In the intermediate range, porosity is introduced by the packing density of cementitious system components, including water. Finally, the finest porosity, measured on the nanometer scale, is associated with the structural development of the hydration products, the so-called "gel" porosity or interlayer porosity.

Control over the porosity and the distribution of pores is achieved both chemically and mechanically. The following discussions will address each of the approaches and factors which affect their performance.

Chemical Control of Pore Distribution

W/C

The single largest factor affecting the porosity and the pore size distribution of any cementitious system is the water-to-cement (W/C) ratio. Minimization of the amount of mixing water is exponentially related to the decrease in porosity[2] and to an increase in strength.[11] For ordinary Portland cement pastes Walton et al.,[2] for example, have found the following relationship:

$$\text{Porosity} = 0.61 + 0.23e^{(w/c)}$$

Chemical Admixtures

A variety of water-soluble organic compounds are routinely used to control the viscosity of concrete and grout mixtures.[12] These materials vary in the magnitude of the effect which they produce and the amounts needed to produce comparable viscosities. The majority are either lignin-based or formaldehyde condensates of melamine or naphthalene sulfonic acid. This latter group is generally available as sodium salts of these acids. The use of these plasticizers and superplasticizers results in mixes with the same viscosity but with W/C ratios that are as much as 20% lower.

Mineral Admixtures

Many different industrial by-products are used in major amounts as reactive mineral admixtures in cements and concretes. Most prominently used are fly ash, granulated blast furnace slag, and silica fume. In particular, the latter two admixtures can be used to significantly alter the porosity and pore distributions of cementitious bodies. Although similar results are achieved with both of these admixtures, their actions are each significantly different. Granulated blast furnace slag (GBFS) is composed of X-ray amorphous aluminosilicate glass

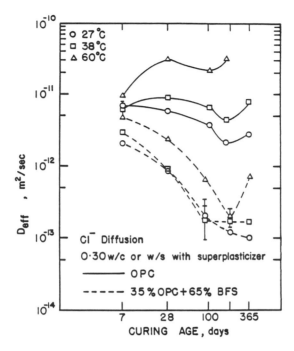

Figure 3-1A.   Effective diffusivity of Cl⁻ ions in OPC vs. slag cements.

which becomes hydraulic (reactive with aqueous phase) upon exposure to alkaline solutions. The hydration of GBFS typically occurs at a slower rate than OPC (ordinary Portland cement; ASTM type I Portland cement) hydration but ultimately achieves comparable physical properties to OPC. A skewing of pore sizes toward the smaller capillary porosity range occurs in this product, directly affecting (reducing) permeability and related diffusivity of ions. This effect is dramatically shown in the thesis work of Kumar[13] and of Kumar and Roy[14] for the diffusivity of Cl⁻ ions through OPC and a 65% GBFS/35% OPC blend, Figure 3-1A. Here the blend exhibits approximately half an order of magnitude lower diffusivity after 7-d curing relative to the baseline OPC data. Although both show a slight decrease in diffusivity as a function of time to 365 d, the blended cement exhibits a dramatic 1.5 orders of magnitude drop. The difference in porosity between the OPC and the blended sample is typically about 5% after 7 d of curing, and increases to approximately 10% after 28 d (the blended is lower), exhibiting only a very slight decrease with increasing temperature, Figure 3-1B.

In a comparable fashion, the very high reactivity of silica fume serves to minimize formation of portlandite, a relatively soluble hydration product of OPC. The submicron particle size of silica fume tends to both maximize packing of particles and also to provide hydration products within the fine porosity, ultimately shifting the distribution to finer sizes.

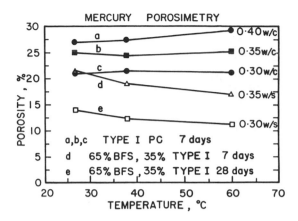

Figure 3-1B.   Total porosity in OPC vs. slag cements with varying W/C.

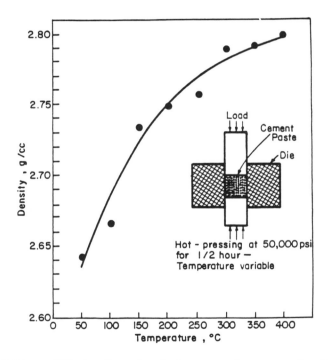

Figure 3-2A.   Effects of mechanical compaction upon density of cements.

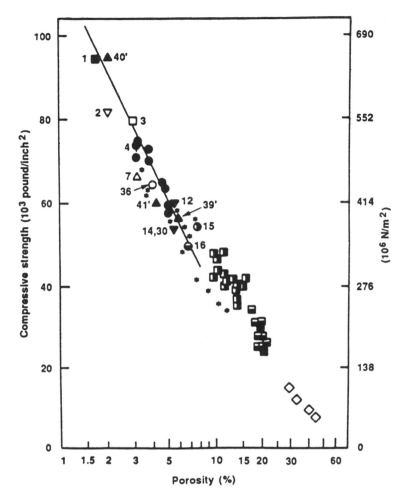

Figure 3-2B. Strength vs. porosity for a variety of cements, ranging from normal (high porosity) to warm pressed (vs. low porosity). (From Roy, D. M. and G. R. Gouda. 1975. *Cem. Concr. Res.* 5(2): 153–162. With permission.)

Mechanical Control over Pore Distribution

Compaction

Roy and Gouda[15] have demonstrated that processing of cement mixes which results in the removal of the bulk of extrinsic porosity yields greater density and higher strength materials with very low porosity, Figure 3-2A and B. This also resulted in greatly reduced solubility/leachability of wasteforms devel-

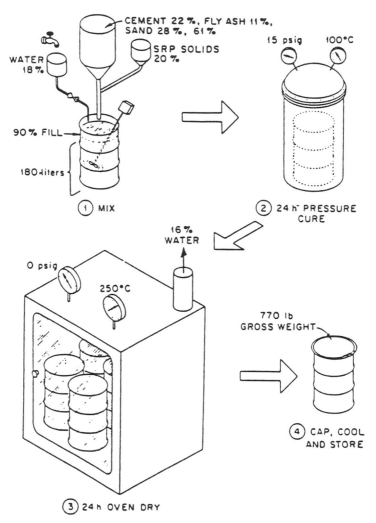

Figure 3-3.   Schematic drawings for FUETAP process.

oped in this manner.[16] The principle of dense packing through pressure application was recognized by Birchall et al.[17] and incorporated into a processing scheme of high shear mixing with a water soluble polymer to form macro defect-free (MDF) cement. The theoretical basis for the improved strength was attributed to the removal of large flaws which act as points of weakness. The mathematical relationship used to describe these observations was the familiar Griffith equation.

On a larger scale, Dole et al.[18] developed a pressure method directly applicable to 96-l drum solidification and compaction. The FUETAP (Formed

Under Elevated Temperature and Pressure) process was designed to process a drum at slightly elevated pressures and temperatures to "epithermally" cure the waste form *in situ*, Figure 3-3. Roy[19] reviewed other processes.

The most likely means of transport for waste ions to the wasteform surface is through the tortuous pathways in the pore structure. Therefore, the mere presence of porosity in a waste form does not by itself suggest a poor product.[13,14] On the contrary, Wise et al.[20] suggest that materials even with relatively large amounts of porosity can be vacuum tight if the porosity is not interconnected. The pore fluids are present, but the ions cannot transport through them, and liquids are not able to move, either because the pores are isolated entities or because they are sufficiently small that the pore fluids and ions (often hydrated) cannot move within them.

Vibratory Densification

A properly placed cementitious body, whether it be a slab of highway concrete or a nuclear wasteform, must be handled and placed with the utmost skill. Vibratory compaction to remove extrinsic porosity introduced during mixing and placement is highly desirable.

Pore Fluid Chemistry

Retention of waste materials within a cement matrix is limited by the transport of the waste ions from the interior to the surface. The most rapid mechanism for this transport is diffusion via the aqueous pore fluids. The chemical nature of these fluids is particularly important in cases where there is a relatively high solute content that is instrumental in solubilizing the immobilized wastes, Figure 3-4.

The following discussion is presented in several parts, most of which are very closely related and are not easily individually separated.

pH Effects on Solubility of Waste

A knowledge of the chemistry of the pore fluids is fundamental to the understanding of the host matrix/waste interaction in any waste form system. For wasteforms fabricated from OPC, the pore fluids will be influenced by the relatively high concentrations of alkalies and to a lesser extent by the alkaline earth components of the hydrating cement phases. Steady-state values of hydroxyl ion concentrations are typically in the range of 13 to 13.5 pH due in major part to alkali species present in the original cements. The concentration of hydroxyl anions is sufficiently large in these systems that the aqueous

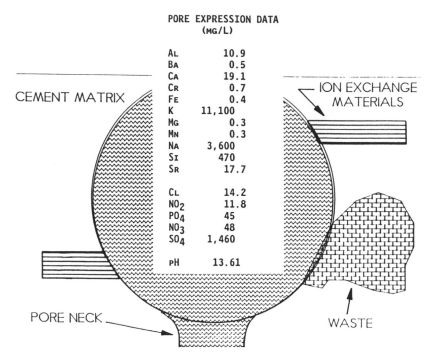

Figure 3-4.   **Schematic drawing of pore environment in normal Portland cement.**

chemistry of the waste ions in question can be addressed as hydroxyl complexes of the general formula:

$$M_x(OH)_y^{(Vx-y)+}$$

where   M = the cation
        V = valence.

The majority of hydroxyl systems of the transition metals and the transuranic elements exhibit solubilities which reach a minimum in the region of near neutrality and increase with both increasing acidity and basicity. The range of minimal solubilities for this class of cations is between 4 and 14, thus offering a range for effective engineering of the wasteform. For example, lead with a minimum solubility at approximately pH 11 will be less effectively immobilized in the hydroxyl form in *normal* Portland cement contrasted to manganese which has its minimum at pH 13.5, the usual pH of OPC pore fluids. Furthermore, the form of the anion will impact the magnitude of

Table 3-1. Dominant Species in Cement Pore Solutions

Cations	Anions	Neutral
Ca^{2+}	Cl^-	$H_4SiO_4^0$
Mg^{2+}	SO_4^{2-}	$H_2CO_3^0$
Na^+	HCO_3^-	
K^+	$Fe(OH)_4^-$	
H^+	$Al(OH)_4^-$	
$MgOH^+$	$H_3SiO_4^-$	
	$H_2SiO_4^{2-}$	
	OH^-	
	CO_3^{2-}	

solubility, lead phosphate being significantly less soluble than lead hydroxides.[21]

E_h Effects on Polyvalent Ions

E_h is defined as the oxidation potential and as such defines the redox state for polyvalent ions. Ions with multiple oxidation states such as trivalent and hexavalent chromium and tetravalent and hexavalent uranium exhibit different solubilities in the different oxidation states. For these two examples cited, the lower valence state is considerably less soluble than the higher state. The presence of finely divided iron filing has been demonstrated to significantly reduce the amounts of soluble mercury in cementitious mixed wastes.[22] Similarly, the use of certain metallurgical slags, such as GBFS, as reactive mineral admixtures, in addition to modifying the pore structures, also contain sufficient amounts of reduced metals or sulfur so that they too can control the redox state of many ions.[23] Here, the electromotive series can be used as a general guide to the selection of proper additives.

Speciation of Ions in Pore Fluids

Table 3-1 lists the dominant aqueous species for the cations and anions in typical pore fluids at pH values of about 13.5. The alkali metals (except for K) and the alkaline earth metals exist as solvated cations, with the exception of the stable hydroxyl-magnesium ion. In contrast, both aluminum and iron are dominantly present as anions. The other species of importance are silica and, subordinately, carbonate, which are present in the form of both anionic and neutral species.

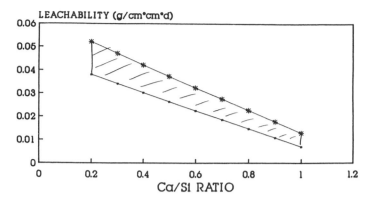

Figure 3-5.   Leachability of cement as a function of Ca/Si ratio.

Table 3-2. Effects of Chemically Adjusting Cementitious Matrix

Adjusting	OPC	ADJ
Lowers pH	11.3	10.8
Lowers Al leaching[a]	3	0.77
Lowers Cs leaching	5.8	0.24

[a] Differential leach rate expressed as grams per square meter per day at 18 d.

1.5. MATRIX STABILITY

Bulk Chemistry Effects

To ensure maximal durability of a cementitious object, the matrix must be designed to exhibit a minimal solubility. For portland cement-based systems cured at ambient conditions, this objective is achieved by controlling the Ca/Si ratio of the anhydrous cement phases and the reactive mineral admixtures. Work conducted by Barnes et al.[28] has demonstrated that a minimum leachability exists for a Ca/Si ratio of 1.0, increasing as the ratio is changed in either direction. Figure 3-5 illustrates this effect for the C/S range of 0 to 1.2 only.

Numerous studies of simulated wasteforms have been conducted in which the host matrix was chemically modified by the additions of reactive silica. The principal effect here is to minimize the formation of portlandite [$Ca(OH)_2$] as a hydration product of alite (Ca_3SiO_5) and belite (Ca_2SiO_4) and to enhance the C-S-H gel formation. Indirectly, this type of modification also can exert a minimal control over pH of the pore fluids. Table 3-2 contains leach rate data expressed as grams per square meter per day for an OPC wasteform, contrasting with the effect of the chemical adjustment to OPC. Typical pH control

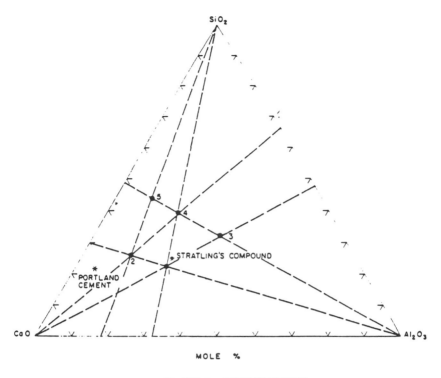

MOLE %

28 DAY Cs CONCENTRATONS*

COMPOSITION	LEACHATE	PORE SOLUTION
2	2.6	75
5	1.3	23

* CONCENTRATIONS IN MMOLE/L

Figure 3-6.   Compositional relationship of bulk cement composition to leachability of Cs⁺.

amounts to lowering by about one half of a pH unit. Dramatic reductions of leach rates for hazardous ions can be achieved, as demonstrated by the approximately 20-fold reduction for Cs⁺ in this study. Hoyle[25] studied a region of compositional space about Stratling's compound in the $CaO-Al_2O_3-SiO_2-H_2O$ system. A sampling of her data is presented in Figure 3-6. The compositions #2 and #5 span the compositional region below and above which, respectively, portlandite is expected to be stable and coexist with the cement hydrate phases. The accompanying chemical analyses of the leachate and the pore fluid compositions clearly demonstrate that reactions with the matrix phases are exerting a significant influence upon Cs retention in this case. The pore fluid compositions are 20- to 30-fold more concentrated with respect to Cs than the leachate, with the pore fluids in the silica-modified wasteforms exhibiting a 3-

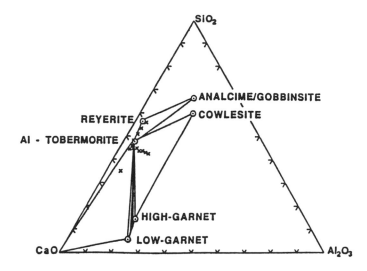

Figure 3-7.   Phase compatibility in CaO-SiO_2-Al_2O_3(-H_2O) system at 175°C. H_2O is the fourth component; CaO is hydrated to $Ca(OH)_2$.

Table 3-3. Selectivity of Al-Tobermorite for Radioactive and Hazardous Cations

	Minimum K_d	Maximum K_d
Cs	100	3,400
Pb	20,000	1,000,000
Cd	500	1,000,000

fold decrease in Cs concentration. Barnes and Scheetz[24] have reported the steady-state compatibility relationships for this system at 170°C, which show the stability field for portlandite, Figure 3-7.

Phase Chemistry

The presence of some materials, either added to the cementitious matrix or formed during hydration, have the effect of exchanging ions between the phase itself and the pore fluids. Data reported by Komarneni et al.[26] have shown that tobermorite formed with maximal aluminum substitution[24] is extremely efficient in exchanging both radioactive and hazardous cations. They report the ranked order of selectivity as being

$$Pb^{2+} > Cd^{2+} > Co^{2+} > Ni^{2+} > Cs^{1+}$$

with typical K_d values reported in Table 3-3.

1.6. CONCLUSIONS

The success of cement-based wasteforms in retaining hazardous ions can be attributable principally to the structure of the porosity of the wasteform and secondarily to the solubility of the species concerned. It is through this avenue that transport from the interior to the surface is most probable, and it is also through this structure that aggressive external fluids and ions can penetrate deeply into the cementitious matrix. Therefore, any control over the porosity and size distribution of this porosity should impact the behavior of the wasteform. Tailoring the bulk chemistry (as well as minimum solubility) of the cementitious matrix to achieve optimum porosity can be achieved with a knowledgeable selection of reactive mineral admixtures based both upon the chemical and physical properties of the admixture. Further, the synergistic effects among properties of the wasteforms must be understood and accounted for if the wasteforms are to perform as anticipated. It is recognized that ideal optimization may not be achieved in bulk applications, but should be approached as much as feasible for each situation.

REFERENCES

1. Atkinson, A. and J. A. Hearne. 1984. "An Assessment of the Long-Term Durability of Concrete in Radioactive Waste Repositories." U.K. Atomic Energy Authority, Harwell (AERE-R 11465).
2. Walton, J. C., L. E. Plansky, and R. W. Smith. 1990. "Models for Estimation of Service Life of Concrete Barriers in Low-Level Radioactive Waste Disposal. Idaho National Engineering Laboratory, Idaho Falls, ID (NUREG/CR-5542, EGG-2597.)
3. Abate, C. 1990. "Synthesis, Characterization, Equilibria and Dissolution Kinetics of Friedels Salt." M.S. thesis, The Pennsylvania State University, University Park, PA.
4. Roy, D. M., M. W. Grutzeck, and K. Mather. 1980. "PSU/WES interlaboratory Comparative Methodology Study of an Experimental Cementitious Repository Seal Material." Office of Nuclear Waste Isolation, Battelle Memorial Institute, Columbus (ONWI-198).
5. Lea, F. M. 1970. *The Chemistry of Cement and Concrete*, 3rd ed. Arnold, Glasgow.

6. Glasser, F. P. 1991. "Long Term Chemical Stability." *Eng. Found. Conf.*, Potosi, MO. In press.

7. Roy, D. M. 1986. "Mechanisms of Cement Paste Degradation Due to Chemical and Physical Factors." Proc. 8th Int. Congr. Cement., Brazil, V: 362380.

8. Lee, J. H., D. M. Roy, P. H. Licastro and B. E. Scheetz. 1991. Nuclear Waste Management IV, Ceramic Transactions, Vol. 23, G. G. Wicks, D. F. Brickford, and L. R. Bunnell, Eds., American Ceramic Society, Columbus, OH. pp. 171–180.

9. Scheetz, B. E., M. R. Silsbee, and A. Kumar. 1987. "Interpretation of Pore Structure in Cementitious Solids by the Use of Fractal Geometry." Oral Presentation, Materials Research Society, Boston.

10. Shi, D., W. Ma, and P. W. Brown. 1990. "Lognormal Simulation of Pore Evolution during Cement and Mortar Hardening." *MRS Proc.*, Vol. 176, Materials Research Society, Pittsburgh. pp. 143–148.

11. Scheetz, B. E., D. M. Roy, C. Tanner, M. W. Barnes, M. W. Grutzeck, and S. D. Atkinson. 1985. "Properties of Cement-Solidified Radioactive Waste Forms with High Levels of Loading." *Ceramic Bulletin* 64(5): 687–690.

12. Malhotra, V. M., Ed. 1989. "Superplasticizers and other Chemical Admixtures in Concrete" Proc. 34th Int. Conf., American Concrete Institute, SP-119, Detroit.

13. Kumar, A. 1985. "Diffusion and Pore Structure Studies in Cementitious Materials." Ph.D. thesis, The Pennsylvania State University, University Park, PA.

14. Kumar, A. and D. M. Roy. 1986. "Pore Structure and Ionic Diffusion in Admixture Blended Portland Cement Systems," Proc. 8th Intl. Congr. Cement Chem., Brazil, V: 73–79.

15. Roy, D. M. and G. R. Gouda. 1975. "Optimization of Strength in Cement Pastes." *Cem. Concr. Res.* 5(2): 153–162.

16. Roy, D. M. and G. R. Gouda. 1978. "High Level Radioactive Waste Incorporation into (Special) Cement." *Nucl. Tech.* 40: 214–219.

17. Birchall, J. D., A. J. Howard, and K. Kendall. 1980. European Patent No. 0055035.

18. Dole, L. R., G. C. Rogers, M. T. Morgan, D. P. Stinton, J. H. Kessler, S. M. Robinson, and J. G. Moore. 1983. "Cement-Based Radioactive Waste Hosts Formed under Elevated Temperature and Pressures (FUETAP Concretes) for Savannah River Plant High-level Defense Waste." Oak Ridge National Laboratory, Oak Ridge, TN (ORNL/TM-8579).

19. Roy, D. M. 1990. "Cementitious Materials in Nuclear Waste Management." *Cements Research Progress 1988.* American Ceramic Society, Columbus, OH pp. 262–292.

20. Wise, S., J. A. Satkowski, B. E. Scheetz, J. M. Rizer, M. L. Mackenzie, and D. D. Double. 1985. The Development of a High Strength Cementitious Tooling/Molding Material." *Very High Strength, Cement-Based Materials.* MRS Proc. Vol. 42, Young, J. F., Ed. Materials Research Society, Pittsburgh. pp. 253–256.

21. Nriagu, J. O. 1978. *The Biogeochemistry of Lead in the Environment, Part A.* (Topics in Environmental Health), Elsevier/North-Holland, New York.

22. Bostick, W. D., J. L. Shoemaker, R. L. Fellows, R. D. Spinel, T. M. Gillian, E. W. McDaniel, and B. S. Evans-Brown. 1988. "Blast Furnace Slag-Cement Blends for the Immobilization of Technetium-Containing Wastes." Martin Marietta, Oak Ridge Gaseous Diffusion Plant (K/QT-203).

23. Malek, R. I. A. and D. M. Roy. 1987. "Stability of Low-Level Cement-Based Waste Systems." Waste Management 87, Am. Nucl. Soc., Tucson, 3: 363–368.

24. Barnes, M. W. and B. E. Scheetz. 1991. The Chemistry of Al-Tobermorite and its Coexisting Phases At 175°C, Vol. 179. Scheetz, B. E., A. G. Landers, I. Odler, and H. Jennings, Eds. Materials Research Society, Pittsburgh. pp. 243–272.

25. Hoyle, S. Q. 1988. "Cesium and Strontium Partitioning during Hydration of Calcium Aluminosilicates." Ph.D. thesis, The Pennsylvania State University, University Park, PA.

26. Komarneni, S., E. Breval, D. M. Roy, and R. Roy. 1988. "Reactions of Calcium Silicates with Metal Cations." *Cement Concrete Res.* 18: 204–220.

27. Roy, D. M. 1987. "New Strong Cement Materials: Chemically Bonded Ceramics." *Science* 235: 651–658.

28. Barnes, M. W., D. M. Roy, and C. A. Langton. 1986. Leaching of Saltstones Containing Fly Ash, in *Fly Ash and Coal Conversion Byproducts: Characterization, Utilization, and Disposal II*, Mat. Res. Soc. Symp. Oral presentation.

4 THE POTENTIAL OF SURFACE CHARACTERIZATION TECHNIQUES FOR CEMENTITIOUS WASTE FORMS

G. N. Salaita and P. G. Hannak

4.1. INTRODUCTION

This review is intended to outline a variety of the modern surface spectroscopic techniques to obtain detailed information on the atomic level of waste materials. These waste materials include untreated wastes, solidified/stabilized wastes, or wastes treated by other means.

Until recently limited research efforts were focused on the microscopic surface characterization of laboratory test samples and sample of feasibility studies. As an auxiliary tool to leachability studies a variety of microcharacterization techniques were used to provide information on the morphology of the solids, changes in surface composition and structure, direct monitoring of accidental contamination, surface redox reactivity, and ion exchange as a function of pH solid waste before and after leaching.

Leaching tests, the essential characterization techniques, give the mean concentration of the contaminants released regardless of their original distribution, which may be quite heterogeneous among a variety of solid phases. Moreover, many leaching tests (e.g., batch test and sequential batch test) give

0-87371-748-1/93/$0.00+.50
© 1993 by Lewis Publishers

only a spatial mean concentration of release or the total concentration over leaching time.

The authors believe that the wide range of waste research, development, and quality control problems can only be solved by a technique or techniques which probes the composition, surface or internal, of a material with both high spatial resolution and elemental sensitivity. Over the past few years research has been started for the chemical characterization of element structure of the solid waste. However, correlation of results from such studies is lacking in the literature.[1-29]

Several excellent modern instrumental techniques are useful to the understanding of the behavior of wastes in the leaching process. These include the four basic surface analysis techniques which are widely employed in all fields of science today: Auger electron spectroscopy (AES), electron spectroscopy for chemical analysis (ESCA), secondary ion mass spectroscopy (SIMS), and ion scattering spectroscopy (ISS).

Development of these techniques occurred in response to the desire to learn about the composition and other properties of material surfaces. With the evolving needs of the waste management, however, the applications were extended to this field. Auger spectroscopy provides localized quantitative surface composition information with high spatial resolution (100 Å). ESCA produces nondestructive quantitative chemical state information of the surface elemental composition. SIMS and ISS yield very high sensitivity and direct chemical information about elemental and molecular surface constituents. Each of these techniques has unique capabilities and problems for which it is the most applicable. However, most studies today require a combination of at least two or three of these techniques and preferably all four.

It is also well recognized that physical aspects such as morphology, roughness, and bulk composition play a role in leaching of waste material. Techniques more restricted to bulk analysis — atomic absorption, inductively coupled plasma, electron dispersive spectroscopy (EDS), etc. or surface physical studies without chemical information such as scanning electron microscopy (SEM) and optical microscopy — will not be discussed here but need to be considered in any complete treatment of waste materials.

Most surface analysis techniques operate under ultra-high vacuum (UHV) conditions due to the mean free path of the electrons and ions used. UHV refers to pressures in the range of 10^{-8} to 10^{-11} torr. The chamber used to maintain these pressures is made of 304 stainless steel. All connections are made by tungsten welding under inert gas blanket or by using Wheeler or Conflat metal seal flanges. Most UHV systems use dual-sorption pumps, cryopumps, turbopumps, diffusion pumps, ion pumps, or a combination of two or three of the above to obtain 10^{-11}-torr pressure. In addition all UHV systems have a liquid nitrogen-cooled titanium-sublimation "getter" pump. The term "ultra-high vacuum" has a special meaning in the context of surface investigation: UHV environment is one in which no significant contamination of the surface occurs.

In any event, cleanliness of the sample is the essential criterion, not system pressure, as the acceptable residual gas pressure for a given experiment depends upon the nature of the species present in the residual gas and upon whether the sample surface is interactive with those species. It is important to note that the application of UHV may lead to some method artifacts. There may be undesirable changes in the physical properties of the subject waste under the high-vacuum conditions. These may lead to desorption, volatilization of some constituents, or to dehydration reactions. Application of a cryogenic system offers the means to diminish these changes.

4.2. X-RAY PHOTOELECTRON SPECTROSCOPY (XPS)

X-ray photoelectron spectroscopy (XPS or ESCA) is one of the many electron spectroscopies particularly useful for the chemical analysis of waste materials. The basic ESCA experiment involves exposure of the sample to be studied to a flux of mono-energetic soft X-rays, ultraviolet light, or synchrotron radiation.

In order to eject electrons from the inner orbital of an atom (the "core" region) most ESCA instruments employ dual anodes of Al K_α and Mg K_α to generate soft X-rays of 1486.6 and 1253.6 eV (K_α), respectively. When ultraviolet radiation of He (21.22 eV) and He II (40.8 eV) is employed for studying the ejected electrons from an outer orbital of an atom (the "valence" region), the technique is refereed to as ultraviolet photoelectron spectroscopy (UPS). Another, less frequently used photon source (very expensive) is the synchrotron radiation source, in the energy range of 10 to 10^5 eV for probing the electronic structure of atoms and molecules.[30] Since energy must be conserved in the photoemission process, the X-ray photon energy, hv, must be partitioned between the kinetic energy of the electrons which are expelled from their atomic bindings, E_k, and the work expended in removing them from the atom, usually referred to as the electron binding energy, E_b (i.e., hv = E_b + E_k). In practice, hv is known, and E_k is measured by the spectrometer, allowing direct determination of the electron binding energy from the difference between hv and E_k.

The sampling depth is set by the energy-dependent mean-free paths (10 to 20 Å for metals, 20 to 30 Å for inorganics, and ~100 Å for organics). The photoemitted electrons are collected and focused by an electrostatic lens into the slit of the electron analyzer (Reference 31 and references therein).

The most widely used analyzer nowadays is the hemispherical analyzer (30 cm in diameter for most commercial instruments). However, Scienta Analytical recently built an analyzer with 60-cm diameter to obtain higher sensitivity at an equivalent resolution (Reference 32 and references therein). A multichannel detector (for enhanced signals) used to resolve energy and position is placed on the focal plane where the different electrons are simultaneously recorded. These events appear on the screen of the computer as a series of lines with different energies. These lines characterize the examined material, providing oxidation state, elemental composition, chemical bonding, and quantitative information. The technique is widely used for detailed surface analytical problem solving of solid materials (for most stable elements in the periodic table, except hydrogen), gases to a pressure of ~1 to 10^5 torr, and free phase liquids and solutions.[31]

Recently ESCA instrumentation has been advanced in three different areas:

1. Multichannel detectors for improved sensitivity — more gain in (S/N) ratio — than the older instrumentation.
2. The X-ray monochromators have been widely used (recently) with ESCA instruments for better energy resolution (line width of 0.44 eV is achieved on the $3d_{5/2}$ photoelectron line of silver metal) with most X-ray monochromators using Al anode compared to a resolution of 0.85 eV for the same Al anode unmonochromatized. To a much lesser extent these monochromators are combined with a rotating anode.
3. ESCA capability is being expanded with the development and the replacement of the cylindrical mirror analyzer (CMA) with the most efficient filter, the hemispherical energy analyzer. This analyzer is the optimum choice for charged particles where high throughput at high energy resolution is required. It is capable of resolving chemical shifts of a few tenths of an electron volt.[32]

Major developments in the hemispherical analyzer took place in the design of the lens-retardation unit. With the advancement of the electron optics design ESCA machines are available with multiple lens units with three or four variable apertures allowing the collection of electrons from small spots ($\simeq 2$ μ) to be analyzed or microscopically viewed.

The best example of the new scanning imaging photoelectron spectrometer is the "ESCAScope" introduced by Vacuum Generators (Reference 33 and references therein). It employs a hemispherical electrostatic analyzer with a multichannel detector which is position sensitive.

A series of lenses are situated between the sample surface and the inlet slit of the analyzer; the lenses have a set of deflection plates inside so that photoelectrons that leave the sample can enter the lens system and then be deflected in a controlled manner into the inlet slit. The lenses make sure that the photoelectrons enter the analyzer properly, and the deflection plates control what part of the sample surface the electrons may be emitted to the analyzer.

Vacuum Generators has designed the optics in such a way that the spots on the surface, only 2 μ in size, can be imaged. Since the voltage on the plates can be ramped, the photoelectron image of a surface can be mapped. Hence, this new photoelectron spectrometer is a true microscope, and it can produce two-dimensional energy-selected photoelectron images of a small well-defined analysis area of the sample surface with high spatial resolution without resorting to scanning.

The analysis area is easily changed by varying the aperture sizes as given in Equation 1.

$$\text{Analysisis area} = \frac{\text{Aperture area}}{\text{Lens magnification}} \qquad (1)$$

Therefore, the new "ESCAScope" can be used both for photoelectron microscopy and photoelectron spectroscopy. Most instrument manufacturers supply ESCA as a scanning microprobe. Image acquisition is gained point by point (i.e., serial acquisition). Its speed usually depends on the sample. It may be slow taking from 15 min to several hours. However, as a microscope, image acquisition is gained for the whole area (i.e., parallel acquisition) in which case it is accomplished in a few minutes. Therefore, ESCA technique offers an efficient technique to study the nuclear waste leaching phenomena.

ESCA offers advantages over the conventional SEM/EDS technique, including high surface sensitivity, physical identification (surface morphology), chemical information (elemental composition, oxidation states, and electronic properties), relatively low beam damage, and reduction of analysis-induced artifacts (for imaging labile ions of Cl^- and Na^+) from leached layers, which either move or desorb under the electron beam used in SEM and EDS.

Quantitation, percent composition, oxidation states, and detailed analysis of specific phases of materials present in Portland cement and cement-based solid waste such as CaO, $CaCO_3$, $Cr(NO_3)_3$, $Ca(OH)_2$, SiO_2, Al_2O_3, Fe_2O_3, K_2O, MgO, and SO_3^{2-} can be thoroughly examined and identified. Accurate analyses of these cement phases are important for studying solidification and stabilization mechanisms.

ESCA can be used as an angle-dependent technique to measure the elemental composition and chemical bonds between atoms not only on the top layer, but between atoms close to or below the surface. This application is very important for multicomponent waste materials. When some elements have a concentration profile within the electron escape depth (surface layers of different composition with a thickness less than the escape depth), the peak amplitude of that element shows a dependence on the exit angle.[31] Since the photoelectrons travel in a straight line from the sample to the detector, the sampling depth is dependent on the sine of the electron take-off angle measured from the plane of the surface. By tilting the sample in front of the analyzer at varying angles from normal to grazing, a high depth resolution between 5 and 100 Å can be attained, allowing elemental composition and chemical state determination within the sample to be made.[34]

When the energy-dependent depth profiling (by alternating analysis and ion milling to remove the previously analyzed layer) is combined with the angle-dependent measurement, it becomes possible to examine thin film structures which are thicker than 1 μ (10,000 Å) with a precise measurement of the interfacial chemistry. This improvement in sensitivity and the ability to measure the interface are important for the application of XPS depth profiling to the analyses of solid waste forms before leaching and after leaching to determine the elemental composition and chemical bonding to different phases of the elements (Cr, Cd, V. Si, Al, Ca, and Fe) with respect to a variety of changes, including pH, concentration, oxidation/reduction, precipitation, and waste form decomposition.

Angle-dependent ESCA is already used widely for nondestructive depth profiling and for determination of uniform overlayer thickness (20 to 100 Å) on substrate. This special branch of angle-dependent ESCA is known as X-ray photoelectron diffraction (XPD).[35] XPD patterns can be used as a fingerprint analysis for comparison between the surface lattice structure of overlayers and those of the underlaying substrate, i.e., epitaxially from oxide on silicon (100) surface forms amorphous phases as SiO_2, or $Ca(OH)_2$, MgO, or CaO. These components are widely encountered in waste materials; therefore, the applicability of the method appears to be feasible. However, it is acknowledged that the methodology was developed and historically applied to single crystal morphology.

Another advantage of ESCA is the simultaneous emission and recording of the photoelectrons with secondary Auger electrons from the X-ray-irradiated sample. Correlation of the Auger electron lines with the photoelectrons can provide very valuable chemical state information. By plotting the kinetic energy of the photoelectron lines of the various chemical states of an element on the abscissa and that of the major Auger line on the ordinate, a two-dimensional chemical state plot is obtained (called the Auger parameter). When standards are used such a plot provides a very powerful tool to define the chemical state in an unknown sample. Since the Auger parameter is the difference between two potential lines in opposite direction, it is charge and reference independent. The Auger parameter can be powerful in determining the chemical environment for a complex situation such as found when leaching nuclear wastes.

4.3. AUGER ELECTRON SPECTROSCOPY (AES) AND SCANNING AUGER MICROSCOPY (SAM)

The field of Auger spectroscopy as applied to chemical problems has evolved like that of ESCA in the last few years. AES is one of many electron spectroscopies particularly useful for studying the elemental composition, packing density, and chemical state of surfaces. It has been employed by one of the authors to study liquid-solid interface of adsorbed organic molecules from aqueous solution at different transition metal surfaces.[36,37]

The sensitivity is about 1% of a monolayer thickness and it is relatively easy to use in comparison with other techniques of electron spectroscopy. All elements except H and He can be detected.

The Auger electron emission process occurs as a collimated electron beam directed against the surface at 2000 to 30,000 eV kinetic energy. Electrons which have binding energies less than the incident beam will leave the inner atomic orbital creating a singly ionized excited atom. The electron vacancy thus formed in the inner shell of an atom can be filled by deexcitation of electrons from other electron energy orbitals. The energy released due to this electronic transition can be either radiated as an X-ray quantum or is transferred to another electron in the same atom or in a different atom. If this electron has less binding energy that the energy transferred to it from the deexcitation, it will be ejected into the vacuum leaving behind a doubly ionized atom. This ejected electron is called an Auger electron.[30,38]

In the last few years scanning Auger microscopy (SAM) has become one of the most powerful and popular surface analysis methods in a wide range of application areas. These areas include ceramics, oxides, glass, microelectronics, particle and contamination identification, composites, metallurgy, polymers, and organics. Several of these areas are directly related to waste materials.

In the last 1 or 2 years, instrument manufacturers have developed high-resolution scanning Auger microprobe systems. These systems operate at a beam energy between 2 and 30 keV by using either the Schottky thermal field emission electron gun with a ZrO/W source tip or the electromagnetic column with a LaB_6 source tip. These guns provide a beam current less than 1 nA into a probe diameter of <150 Å.

A detection system based on a cylindrical mirror analyzer and the hemispherical energy analyzer with a multichannel detector has been used to provide high sensitivity and to reduce sample damage. These Auger systems increase the beam current in a small diameter beam by a factor of 20 to 100, and the multichannel detector increases the sensitivity by a factor of 10, resulting in an approximate 200 to 1000 increase in the overall instrument speed. The very intense small electron beam generated by the above-mentioned sources helped to incorporate the SEM/SAM into one instrument.

This newly emerged SAM system is a very useful and powerful tool in identification and quantitation of the surface characteristics of waste materials in terms of physical and chemical properties.

The first step in the Auger analysis is the investigation of surface morphology and topography by SEM. It is then followed by the identification of areas of interest on the surface and a quantitative elemental and chemical analysis using point or area Auger electron spectroscopy (AES). To produce Auger spectra, multi-dimensional analysis may then be carried out using SAM line scans and maps. These give a wealth of information on the spatial distribution of elements on the surface which may be compared directly with SEM.

In addition, this technique can be used as a destructive sputter profiling to investigate elemental and chemical distributions to a depth of 10,000 Å (by using alternating Auger analysis and ion beam sputtering to remove the previously analyzed layer). Because of its inherently higher S/N ratio, AES is the preferred technique for depth profiling. This approach is useful in monitoring the removal of successive layers of materials (as in the case for leached wastes), in fiber/matrix investigations, and in ceramic thin film/substrate interface studies. Oxide layer thickness and depletion zones can be identified as in cementitious matrix.

In summary, Auger spectroscopy can be used to provide point analysis of specific sites, or multiple-point Auger analysis can be used on different parts of the material surface (if it is a dark area or light area as first looked at by SEM). The chemical composition of the dark area and the chemical composition of the light area can be searched, This allows the experimenter to examine the segregation of the different waste phases or adsorbate orientations (as a function of concentration, temperature, pH, electrolyte, solvent, surface roughness, and traces of surface-active impurities) formed on the surface relative to the bulk.

4.4. SECONDARY ION MASS SPECTROSCOPY (SIMS)

SIMS

Secondary ion mass spectroscopy (SIMS) is carried out by ion sputtering (or ion etching) the sample with primary ions, usually argon, oxygen, or cesium ions, having an energy from a few hundred up to several thousand electron-volts. Sputtering species are emitted as neutrals in various excited states, as ions both positive and negative, singly and multiply charged, and as clusters of particles.[39]

Analysis of sputtered species is the most sensitive of the surface analysis techniques. It can detect minor glass components at low concentrations (<1 ppm) and it has the ability to detect hydrogen.[40] The ions sputtered by such a process are called secondary ions as opposed to the primary ions of the probe beam, and are generally of low energy, 5 to 50 eV.[39-41]

These secondary ions, representative of the composition of the outer-most atomic layers of the specimen, can be extracted into a mass spectrometer for mass-to-charge ratio separation and detection, thereby providing a qualitative and quantitative analysis of the surface region (one to three monolayers). Thus, SIMS analyzes the material removed from the sample surface by sputtering, in contrast to the Auger spectroscopy which analyzes only the outer few atomic layers present on the surface without material removal.

Since SIMS involves analysis of ions formed on the surface, it is especially sensitive to elements that are easily ionized, i.e., low ionization potentials. It is extremely sensitive (reaching the level of parts per million or even parts per billion) to strongly electropositive elements such as Li, Na, K, Be, Mg, Ca, Ba, etc., and in general is highest in detection sensitivity to elements with a mass less than 120. SIMS also becomes very sensitive to electronegative elements such as C, N, O, F, Cl, S, etc.

SIMS may be divided into two areas: static and dynamic. Static SIMS is a surface analysis technique; dynamic SIMS is a depth analysis technique. There are no specific boundaries for the two techniques and the classifying parameter is the primary beam current density.

Typical current densities in static SIMS are on the order of 10 nA/cm^2. Assuming a density of 2×10^{15} atoms per square centimeter at the sample surface and a yield of 1:1, 10% of the top layer will be removed in 1 h. This means that only ions from the top-most layer are being acquired during a normal experiment.

For dynamic SIMS the primary ion beam intensity is of the order of 1 to 10 mA/cm^2 and is rastered across the surface to produce a flat-bottomed crater. The SIMS signal is reflected from the center of the crater plotted against time or depth for various elements giving a depth profile analysis.

In ion sputtering depth profiling, electronic secondary ion gating (the raster scan technique is necessary to achieve a high dynamic range and depth resolution. Without electronic gating, secondary ions, which originate near the edges, are different in concentration and depth than those ions from the center of the crater.

The uniform erosion of the sample, i.e., in the form of a flat-bottomed crater, is absolutely crucial since it determines the depth resolution of the measured concentration profile. Thus, the measured depth profile is constructed from only those secondary ions which originate in the central region of the crater area.[41-44] Generally speaking, the region from which ions are collected is about 10 to 16% of the total crater area. The ability to uniformly erode that central area and thereby maintain a flat bottom has been resolved by Perkin-Elmer-Physical Electrons Division (Φ).

Φ has introduced the Zalar Rotation profile. The Zalar rotation allows the sample to rotate on its own axis while being sputtered simultaneously inhibiting cone formation and preferential sputtering. It also allows the generation of a flat-bottomed crater, offering at the same time multiple analyses of the same sample (up to 20 regions selectable, continuous or alternating sputtering).[45] The ability to perform multiple analyses on the same sample is beneficial where unstable charging requires that the analysis be repeatedly terminated and reinitiated. Also, it helps to minimize the instrumental problem which is of concern in depth profiling of leached nuclear waste glass surfaces relative to spectral resolution and detection sensitivity.

Another technique that has been reported by Φ to allow the formation of a flat-bottomed crater besides the Zalar rotation and the raster scan technique is the spiral scan technique.[44] This spiral scan rate is faster than the raster scan so that the secondary ion will reach the detector when the primary ion beam is actually in the center of the crater.[44]

Imaging SIMS

The advent of high-brightness liquid metal ion gun has moved SIMS into the realm of surface microanalysis.[41] Because of their high brightness, the liquid metal ion sources have resulted in focused ion-beam systems with submicron spot sizes and current densities in excess of 1 A/cm^2. The use of these sources has led to an improvement in the lateral spatial resolution of SIMS. Systems with spatial resolution below 500 Å are now available commercially. These sources are typically metals that are liquid near room temperature, e.g., gallium, indium, and cesium. However, gallium is the most popular source material. It has a melting point of 28°C, a vapor pressure below 10^{-9} mbar, and it is available in high purity form, since it possesses only two intense isotopes causing minimal interference in the SIMS spectrum.

There are two methods of operation. In the first, the liquid metal flows through a fine aperture under the influence of a high electric field relative to its surroundings, forming a cone from the apex of which intense positive-ion emission occurs.

Using the second method, the liquid metal wets and flows over a fine needle. The needle normally is fabricated from tungsten and the radius at the apex usually is about 5 μ. A high positive field in set between the needle and the extractor, and intense jet-like emissions of liquid metal ions occur. These ions are emitted by field evaporation from an extremely small area. These liquid metal ion sources provide microfocusing capabilities achieving 500-Å lateral resolution at 25-keV beam energy. Currents range from 1 pA to 100 nA with continuously variable energy between 0.5 to 25 keV. In addition, these sources are extremely stable and UHV compatible. Two- and three-dimensional imaging, microanalysis (mapping), and small area depth profiting can be achieved simultaneously. This indicates the ability of SIMS to analyze waste materials of widely differing electrical and chemical properties.

Chemical mapping using SIMS can be a particularly powerful technique for studying transport mechanisms of how oxides are formed in nuclear reactors. The ability of SIMS to map both elemental and molecular features (such as composites with fibers and matrix having different electrical conductivities) on a surface is a major advancement in surface chemical microscopy.

Charging

SIMS analysis of insulators has a wide range of applications in waste materials, e.g., glass, organics, and waste metal oxides. Positive ion bombardment of an insulator sample can produced charged layers on the sample surface which modify the secondary ion distribution and make SIMS analysis impossible. Electron bombardment as in AES and photo bombardment as in XPS result in surface charging. Two techniques are available to prevent charging:

(1) neutralization with a low-energy electron beam or (2) replacement of the primary ion beam by a fast atom beam (FAB). The first technique is the most popular and the least expensive.

Secondary Neutral Mass Spectrometry (SNMS)

This system was made by Leybold-Heraeus and was first introduced into the market in 1987. SNMS is closely related to SIMS in that the sample surface is bombarded with noble gas (usually argon) and surface atoms are ejected into a vacuum, collected by a lens system, and then analyzed with a mass spectrometer.

In SIMS only sputtered ions that are not neutralized when escaping from the sample surface are detected; since there are roughly four to five orders of magnitude in the spread of ion neutralization probability among ionic species and radicals, SIMS has been difficult to make quantitative on samples that are of truly unknown composition.

In SNMS all ions and neutrals coming off a surface pass through either a plasma, a laser, or an electron-ionization region where the neutrals can become ionized independent of the species. Thus, variation in surface neutralization is greatly reduced; consequently, the number of ions of an element reaching the mass spectrometer is directly proportional to its atomic concentration in the sample surface region. The capability of the SNMS to capture neutrals is what makes the technique quantitative.

A complied extensive list of sensitivity factors by many instrument manufacturers for most elements has shown SNMS to be as quantitative a caliber as ESCA.

A major disadvantage of SNMS over ESCA is that it is sensitive down to parts per billion levels in conductive samples and to about 10 ppm in insulators.

The disadvantages of SNMS are (1) it destroys the sample by sputtering, (2) it detects elements and small molecular fragments so that few identifications of oxidation states can be made, and (3) it is less surface sensitive than ESCA. It only analyzes up to 10 nm (100 Å) at a time.[42]

Examples related to nuclear waste material studied by SIMS have been presented from the work of Sparrow and Patano (with the authors' permission); the waste form used is glass. Since glass is a supercooled liquid which flows at room temperature, its surface composition would change with heat treatment. In Figure 4-1, the compositional depth profile of heated glass vs. bulk glass is shown. The silicon concentration of the glass fiber increased nearly 50% over the bulk values, whereas the Al, Ca, Mg, and B decreased. These results indicate that heat treatment resulted in substantial compositional changes at least 500 Å into the surface. Shown in Figure 4-2 are SIMS profiles of a borosilicate glass designed to immobilize ICPP zirconium waste calcine.

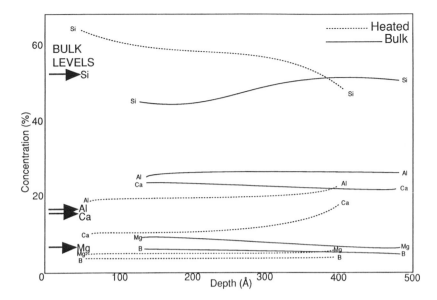

Figure 4-1.   **Semiquantitative results obtained from SIMS depth profile of glass fiber and bulk glass.**

4.5. ION SCATTERING SPECTROSCOPY (ISS)

ISS is the simplest of all surface analysis techniques[44] in principle and application. A monoenergetic beam of positive ions such as He^{3+}, He^{4+}, Ne^+, or Ar^{4+} is directed at the sample. Some of these positive ions are reflected (scattered) with the loss of energy corresponding to the simple binary elastic collision of the beam with a particular surface atom. At any particular angle, the energy loss is only dependent upon the mass of the surface atoms causing scattering since each element has a different mass. Each surface element will cause a different energy loss according to the equation

$$\frac{E}{E_o} = \left(\frac{m}{M+m}\right)^2 \left(\left[\frac{M^2}{m^2} - \sin^2\Theta\right]^{1/2} + \cos\Theta\right)^2$$

where   m = incident ion mass
         M = incident atom mass
         E = scattered ion energy
         E_o = primary beam energy

As ISS spectrum is easily obtained by recording the number of scattered primary ions collected per second as a function of their energy from zero to the energy of the primary beam. The technique is very sensitive to the top 0 to 2

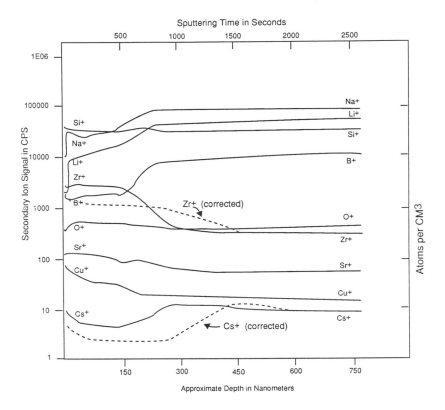

Figure 4-2.   **SIMS depth profiles for a Soxhlet leached nuclear waste glass. The Zr+ and Cs+ profiles were corrected assuming that the leached layer sputters two times faster than the bulk glass (7 keV Ar+ at 0.6 mA/cm².).**

Å of the surface and it is a complementary technique to the SIMS methodology. It is nearly independent of sample matrix effects. Thus, it presents a relatively quantitative indication of the surface concentrations. Detection limits on the order of a few parts per million can be achieved for high mass matrices such as As or Sb in SiO_2.[44]

Depicted in Figure 4-3 is an ISS depth profile analysis of soda lime glass (common window glass). This spectra is reproduced with the permission of Committee E-42 on Surface Science Analysis: American Society for Testing and Materials (STP643) Cleveland, March 1977, McIntyre, N. S., Ed.

Shown in Figure 4-4 are the relative variations within the top 200 Å. Also shown in Figures 4-5 and 4-6 is the cleaning of the glassware with chromic acid or alcohol KOH to remove impurities and prevent further contamination. ISS shows that when laboratory glassware is washed with chromic acid it depletes the surface of alkali and alkaline earth metals leaving the surface essentially as silica (SiO_2).

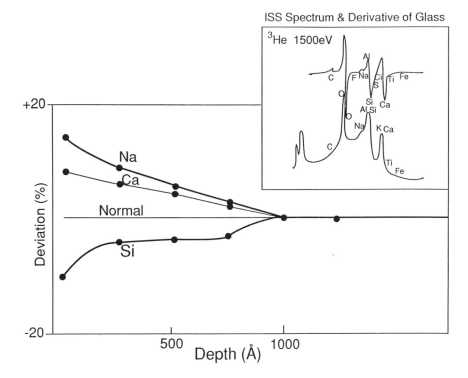

Figure 4-3.   Quantitative ISS illustrating changes of glass surface composition resulting from zirconium polishing. Insert shows a
typical ISS spectrum and its derivative.

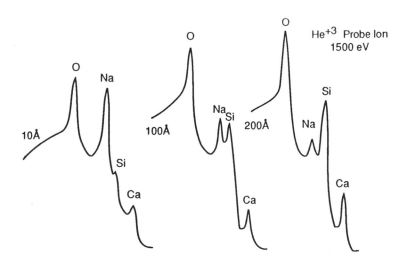

Figure 4-4.   ISS spectra of soda lime glass indicating sodium enrich-
ment at the surface.

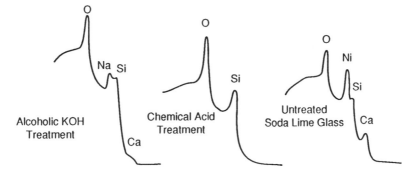

Figure 4-5. Changes in surface composition of soda lime glass as a result of common chemical treatments.

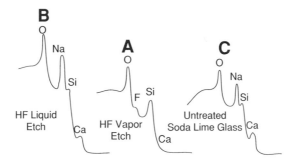

Figure 4-6. Surface changes of glass treated by HF liquid and HF vapor.

This overview gave a cross section of several used and emerging surface characterization techniques. With the application of these techniques a new dimension of waste characterization has opened up. It is essential to understand the capabilities and limits of each of these techniques in order to effectively utilize them. It is anticipated that a future improved communication between researchers of waste treatment and instrument experts would lead to a better understanding of the waste containment and release.

REFERENCES

1. Bishop, P., D. L. Gress, and J. A. Olofsson. 1982. *Proc. 14th Mid-Atlantic Industrial Waste Conference.* Alleman, J. E. and Kavanagh, J. T., Eds. Ann Arbor Science Publishers, Ann Arbor, MI. pp. 459–467.
2. Bishop, P., S. B. Ransom, and D. L. Gress. 1983. *Proc. Industrial Waste Conference,* Purdue University, West Lafayette, IN. pp. 395–401.
3. Bishop, P. 1988. *Haz. Waste Haz. Materials* 5: 129–143.

4. de Groot, G. J., J. Wijkstra, D. Hoede, and H. A. van der Sloot. 1989. *Environmental Aspects of Stabilization and Solidification of Hazardous and Radioactive Wastes.* ASTM STP 1033, Cote, P. L. and T. M. Gilliam, Eds. American Society for Testing and Materials, Philadelphia. pp. 170–183.

5. Poon, C. S., A. I. Clark, C. J. Peters, and R. Perry. 1985. *Waste Management Res.* 3: 127–142.

6. Poon, C. S. and R. Perry. 1987. "Studies of Zinc." *Materials Res. Soc. Symp. Proc.* 86: 67–76.

7. Ivey, D. G., M. Neuwirth, S. Shumborski, D. Conrad, R. J. Mikula, W. W. Lam, and R. B. Heimann. 1990. "Electron Microscopy of Heavy Metal Waste in Cement Matrices." *J. Materials Sci.* 25: 5055–5062.

8. Chick, L. A., G. L. McVay, G. B. Mellinger, and F. P. Roberts. 1980. "Annual Report on the Development and Characterization of Solidified Forms for Nuclear Wastes, 1979." Pacific Northwest Laboratory, Richland, WA (PNL-3465/UC-701).

9. Chickalla, T. D. and J. A. Powell. 1981. "Nuclear Waste Management Quarterly Progress Report, Oct–Dec 1980. Pacific Northwest Laboratory, Richland, WA (PNL-3000-8/4C-70).

10. Clark, D. E. and C. A. Maurer. 1982. "Waste Glass/Repository Interaction" in *Scientific Basis for Nuclear Waste Management,* Vol 5. Lutze, W., Ed. Plenum, New York. pp. 71–82.

11. Clark, D. E., L. Urwongse, and C. Maurer. 1982. "Application of Glass Corrosion Concepts to Nuclear Waste Immobilization." *Nuclear Tech.* 56: 212–225.

12. Clarke, D. E. and L. L. Hench. 1982. "Theory of Corrosion of Borosilicate Glass Corrosion." Paper given at the 6th Int. Symp. Scientific Basis for Nuclear Waste Management, Nov. 1–4, Boston, to be published in the symposium proceedings.

13. Fletcher, W. W. 1974. "The Chemical Durability of Glass." Progress Report in Burial Experiments at Ballidon and Wareham, England.

14. Fox, P. G. 1980. "Thermodynamic Stability of Oxides in Aqueous Solution and its Relevance to Static Fatigue in Silicate Glasses." *Phys. Chem Glasses* 21(5): 161–166.

15. Freeborn, W. P., M. Zoelensky, B. E. Sheetz, S. Komarneni, G. J. McCarthy, and W. B. White. 1982. "Hydrothermal Interaction between Calcine, Glass, Spent Fuel, and Ceramic Waste, Forms with Representative Shale Repository Rocks." Materials Research Laboratory, Pennsylvania State University, University Park, PA.

16. Hall, A. R., A. Hough, and J. A. C. Marples. 1982. "Leaching of Vitrified High Level Waste" in *Scientific Basis for Nuclear Waste Management,* Vol. 5. Lutze, W., Ed. Plenum, New York. pp. 319–328.

17. Hayward, P. J., W. H. Hocking, F. E. Doern, and E. V. Cecchetto. 1982. "SIMS Depth Profiling Studies of Sphere Based Ceramics and Glass Ceramics Leached in Synthetic Groundwater" in *Scientific Basis for Nuclear Waste Management,* Vol. 5. Lutze, W., Ed. Plenum, New York. pp. 319–328.

18. Hench, L. L., L. Werme, and A. Lodding. 1982. "Burial Effects on Nuclear Waste Glass" in *Scientific Basis for Nuclear Waste Management,* Vol 5. Lutze, W., Ed. Plenum, New York. pp. 71–82.

19. Hermansson, H. P., H. Christensen, D. Clark, and L. Werme. 1982. "Effects of Solution Chemistry and Atmosphere on Leaching of Alkali Borosilicate Glass." Paper given at the 6th Int. Symp. Scientific Basis for Nuclear Waste Management, Nov. 1–4, Boston, to be published in the symposium proceedings.

20. "Materials Characterization Canter (MCC) Test Methods." Preliminary Version. Pacific Northwest Laboratory, Richland, WA (PNL-3940).

21. Merritt, W. F. and P. J. Parsons. 1964. "The Safe Burial of High Level Fission Product Solutions Incorporated into Glass." *Health Physics* 10: 664–664.

22. Merritt, W. F. 1977. "High Level Waste Glass: Field Leach Test." *Nuclear Technol.* 32: 88–91.

23. Paul, A. and M. S. Aman. 1978. "The Relative Influences of Al_2O_3 and Fe_2O_3 on the Chemical Durability of Silicate Glasses at Different pH Values." *J. Materials Sci.* 13: 1499–1502.

24. Ross, W. A. and J. E. Mendel. 1979. "Annual Report on the Development and Characterization of Solidified Forms for High-Level Wastes, 1978." Pacific Northwest Laboratory, Richland, WA (PNL-3060/UC-70).

25. Savage, D. and J. E. Robbins. 1982. "The Interaction of Borosilicate Glass and Granodiarite at 100°C, 50 MPa; Implications for Models of Radionuclide Release" in *Scientific Basis for Nuclear Waste Management,* Vol. 5. Lutze, W., Ed. Plenum, New York. pp. 145–152.

26. Solomak, A. B. and L. Zumwalt, Jr. 1980. "HLW Fixation in Sintered Modified Synroc-B Ceramics: Chemical Stability Evaluation." *Trans. Am. Nuclear Soc.* 35.

27. VanIseghem, P., W. Timmermans, and R. deBatist. 1982. "Chemical Stability of Simulated HLW Forms in Contact with Clay Media" in *Scientific Basis for Nuclear Waste Management,* Vol. 5. Lutze, W., Ed. Plenum, New York. pp. 219–227.

28. Werme, L., L. L. Hench, and A. Lodding. 1982. "Effect of Overpack Materials on Glass Leaching in Geological Burial" in *Scientific Basis for Nuclear Waste Management,* Vol. 5. Lutze, W., Ed. Plenum, New York.

29. Wicks, G. G., B. M. Robnett, and W. D. Rankin. 1982. "Chemical Durability of Glass Containing SRP Waste Leachability Characteristics, Protective Layer Formation, and Repository System Interactions" in *Scientific Basis for Nuclear Waste Management*, Vol. 5. Lutze, W., Ed. Plenum, New York. pp. 15–24.

30. Somorjai, G. A. 1981. *Chemistry in Two Dimensions: Surfaces.* Cornell University Press, Ithaca, NY.

31. Siegbahn, K. 1990. *J. Electron Spectros. Relat. Phenomena* 51: 11–36.

32. Gardella, J. A. 1989. *Anal. Chem.* 61(9): 589A.

33. Coxon, P., J. Krizek, M. Humpherson, and R. M. Wardell. 1990. *J. Electron Spectros. Relat. Phenomena* 51: 39.

34. Sobol, P. E. 1989. Perkin-Elmer Physical Electronics, Technical Bulletin No. 2, Eden Prairie, MN.

35. (a) Siegbahn, K., C. N. Nordling, A. Fahlman, R. Nordberg, K. Hamrin, J. Hedman, G. Johansson, T. Bermark, S. E. Karlsson, I. Lindgren, and B. Lindberg. 1962. *ESCA: Atomic, Molecular and Solid State Structure Studied by Means of Electron Spectroscopy.* Almqvist and Wiksells, Uppsala.

     (b) Seah, M. P. and G. C. Smith. 1988. *Surface Interface Anal.* 11: 69.

     (c) Siegemund, O. H. W. et al. 1982. *IEEE Nuclear Science Symp.*

     (d) Coxon, P. et al. 1990. *J. Electron Spectros. Relat. Phenomena* 51.

     (e) Gardella, J. A. 1989. *Anal. Chem.* 61(9): 589A.

36. (a) Salaita, G. N. et al. 1988. *J. Electronal Chem.* 245: 253.

     (b) Salaita, G. N. et al. 1987. *J. Electronal Chem.* 229: 1–17.

37. (a) Salaita, G. N. and A. T. Hubbard. 1992. *Catalysis Today.* 12: 465–479.

     (b) Wiechowski, A., S. D. Rosasco, G. N. Salaita, A. T. Hubbard, B. E. Bent, F. Zaera, D. Godbey, and G. A. Somorjai. 1988. *J. Am. Chem. Soc.* 107: 5910.

38. (a) Chang, C. C. 1971. *Surface Sci.* 25: 53–79.

     (b) Harris, L. A. 1974. *J. Vac. Sci. Technol.* 11: 23–28.

     (c) Perkin-Elmer Physical Electronics. 1990. "pHI Model 670 Auger Nanoprobe" technical Bulletin.

39. Feldman, L. C. and J. W. Mayer. 1986. *Fundamentals of Surface and Thin Film Analysis.* North-Holland, New York.

40. Pantano, C. G. 1984. Private communication.

41. (a) Cuomo, J. J., R. J. Gambino, J. M. E. Harper, and J. D. Kuptis. 1978. "Significance of Negative Ion Formation in Sputtering and SIMS Analysis." *J. Vac. Sci. Technol.* 15: 281–287.

     (b) Simons, D. S., J. E. Baker, and C. A. Evans, Jr. 1976. "An Evaluation of the Local Thermal Equilibrium Model for Quantitative SIMS Analysis." *Anal. Chem.* 48: 1341–1348.

(c)   Deline, V. R., W. Katz, C. A. Evans, Jr., and P. Williams. 1978. "Mechanism of the SIMS Matrix Effect." *Appl. Phys. Lett.* 33: 832–835.

(d)   Setti, R. L., J. M. Chabala, and Y. L. Wang. 1988. *Ultramicroscopy* 24: 97–114.

42.   Clark, R. J. H. and R. E. Hester. 1988. *Spectroscopy of Surfaces.* John Wiley & Sons, New York.

43.   Smith, D. P. 1976. *J. Appl. Phys.* 38: 340–347.

44.   Sparrow, G. R. 1977. "Application of ISS/SIMS in Characterizing Thin Layers (10 nm) of Surface Contaminants." Society of Manufacturing Engineers, Cleveland, OH.

45.   (a)   Perkin-Elmer Physical Electronics. 1989. "pHI Technical Bulletin." p. 8905.

(b)   Zalar, A. 1985. *Thin Solid Films* 124: 223.

(c)   Perkin-Elmer Physical Electronics Division. 1990 "pHI Technical Bulletin. p. 9002.

5 ELECTRON MICROSCOPY CHARACTERIZATION TECHNIQUES FOR CEMENT SOLIDIFIED/STABILIZED METAL WASTES

D. G. Ivey, M. Neuwirth, D. Conrad, R. J. Mikula,
W. W. Lam, and R. B. Heimann

5.1. ABSTRACT

Scanning electron microscopy (SEM), transmission electron microscopy (TEM), and scanning transmission electron microscopy (STEM) are shown to be useful in characterizing heavy metal uptake in cement stabilized metal wastes. Examples of these techniques, applied to cement systems, illustrate the advantages and disadvantages and the complementary nature of SEM and TEM. Regions in the cement matrix can be probed by SEM techniques such as backscattered electron imaging, X-ray mapping, and line scanning, as well as conventional imaging and X-ray acquisition. TEM/STEM, on the other other hand, can probe discrete cement phases providing both crystallographic and chemical information. A novel TEM/STEM specimen preparation technique is discussed, along with preliminary results on the mechanism of Cr^{3+} containment in ordinary Portland cement (OPC).

5.2 INTRODUCTION

Electron microscopy techniques, i.e., SEM and TEM/STEM, are quite useful in studying the complex microstructures associated with cement-stabilized heavy metal waste. SEM and TEM/STEM are complementary techniques in the sense that SEM provides relatively large-scale microstructural and compositional information, while TEM/STEM gives much higher resolution or more localized information. Morphological changes in cement associated with the addition of a particular waste, as well as the location of contaminants within the various cement phases, can therefore be monitored through the various imaging, electron diffraction, and X-ray spectroscopy techniques associated with SEM and TEM/STEM. This information can ultimately lead to a better understanding of the stabilization mechanisms.

SEM is utilized in the study of hazardous wastes in cement in a variety of ways from characterization of cement morphology to quantitative elemental analyses. Characterizing gross morphological changes which occur upon the addition of waste metals to cement can be important since the suppression of certain cement phases impacts on strength as well as leaching behavior.

The X-rays excited by the electron beam can provide information about the disposition of certain elements in the cement matrix and (to some extent) their association with particular cement phases. In general, the resolution of this SEM technique is limited to the observation of regions on the order of 1 μm, in terms of the X-ray resolution, in the cement sample.[1-3] TEM, on the other hand, offers greater X-ray resolution (<30 nm) and can probe the discrete associations between metals contained in the cement matrix and particular cement phases.[4-9]

Image resolution in the TEM is also superior (0.2 to 0.3 nm) to that in the SEM (3 to 6 nm), although this is often more than offset by the problems associated with preparing adequate specimens for TEM examination. Electron transparent regions are required for TEM, which means thinning a specimen down to a thickness of about 100 nm. This is especially a problem with cement materials because of their porous and brittle nature. Several techniques have been reported in the literature,[4,5,9] including grinding, extraction, and ion milling. There are many drawbacks associated with these techniques and the resultant specimens are often less than ideal. For example, both grinding and extraction techniques only permit the observation of selected phases at best and not the overall microstructure, while grinding also introduces the additional problem of mechanical damage. Ion milling is the best technique in terms of the finished product, but specimen preparation by this method is very tedious and time consuming. Since specimens can only be mechanically polished to a thickness of 30 to 50 μm, 15 to 30 h of ion milling time is required. Often, even after this procedure has been followed, a high-voltage (1000 kV) electron microscope (HVEM) is required to provide adequate electron penetration for observation.

In this paper, the capabilities and limitations of both SEM and TEM/STEM, with respect to microstructural analysis of cementitious materials, are briefly reviewed. Also, a novel technique for preparing TEM specimens from cementitious materials is described. Results of work done to characterize Cr^{3+} containment in OPC are also presented.

5.3. ELECTRON MICROSCOPY TECHNIQUES

Scanning Electron Microscopy (SEM)

SEM Specimen Preparation

Samples were prepared for SEM observation by fracturing and carbon coating. The carbon coating is required in order to prevent sample charging effects and to minimize X-ray attenuation for the acquisition of compositional information. In certain cases, it is possible to observe the samples uncoated. Gold or other metal coatings are more efficient for the production of micrographs but they attenuate the X-rays to a much greater extent and were therefore not used in this study. The fracturing generally produces an angled surface which must then be tilted during observation in order to present a reasonably flat surface perpendicular to the electron beam. The fracturing was done at room temperature and found not to be observably different in the electron microscope from fractures done at liquid nitrogen temperature.

SEM Examination

The scanning electron microscope used to obtain the results discussed here was a Hitachi X-650 equipped with both wavelength and energy-dispersive spectrometers. An energy dispersive spectrometer, with a 30-mm^3 Si(Li) detector, was used to produce all of the X-ray spectra shown in the figures. The incident beam current was approximately 0.2 nA at 25 keV unless otherwise specified. In order to more clearly illustrate the presence of some of the less-abundant elements, the X-ray spectra are presented on a logarithmic scale. The X-ray emission which is excited by the incident electron beam is characteristic of each of the elements present in the sample. By focusing the incident electron beam it is possible to determine compositional differences as a function of position in the sample. Ideally, with a flat sample surface perpendicular to the incident electron beam, this X-ray information can be made quantitative. With the cement samples used in this study, however, there is always a certain surface "roughness" which makes quantification difficult, especially when examining micron-sized occlusions or concentrations of impurities.

Attenuation of the X-rays from deeper regions in the sample also affects the accuracy of the X-ray analysis. This outlines the importance of specifying electron beam energies, currents, and X-ray acquisition times. The beam energy affects the depth of sample penetration by the incident electrons and therefore the X-ray resolution, while the electron current and acquisition times indicate the extent of electron beam damage to the sample which may be taking place. For samples with low waste metal concentrations, long acquisition times are required to see evidence of the waste metal in the X-ray spectrum. At higher concentrations it becomes possible to identify localizations of waste metals in certain regions of the cement sample. The resolution limitation of the X-ray information means that it is not always possible to determine whether the waste metals are replacing other ions in certain cement phases or precipitating out as inclusions.

By using the appropriate detector, the electron microscope can be used to determine relative elemental or mineral abundances which make it possible to identify materials of different composition. For instance, a BSE (backscattered electron) detector will produce a brighter image from regions composed of higher average atomic number (Z) elements. Therefore, concentrations of the higher Z metals such as cadmium, chromium, and lead will appear brighter than the calcium, silicon, and aluminum in the cement matrix.[10] By correlating the observed brightness to a particular compound, it is then possible to qualitatively determine the relative amounts of heavy metals in a given cement sample. These data can be collected and quantified over a large number of fields of view by interfacing an image analyzer with the electron microscope.[10]

A more direct determination of impurity inclusions can be made by acquiring X-ray dot maps of elements of interest. This involves rastering the electron beam across the sample and correlating X-ray emissions with the position of the electron beam. This can provide more accurate compositional information than the indirect method of measuring the brightness in a BSE detector, especially in the case of cement samples where BSE image brightness can be affected by surface roughness as well as by compositional differences.

Many of the above techniques are demonstrated in the following example of an OPC sample containing relatively low levels of Cr and Cd impurities. This sample was prepared by dissolving salts of the contaminants in distilled water and then treating with OPC. Figure 5-1 shows a conventional SEM micrograph of a sample with two areas marked for X-ray analysis. The spectra corresponding to these areas are shown on a logarithmic scale in the same figure. Although the X-ray spectra clearly show that region B has a concentration of cadmium and chromium, it is not possible to determine that these areas represent different chemical compositions from their morphology alone. However, the BSE image in Figure 5-2 clearly shows a brighter image in the region containing heavy metals (i.e., the higher Z region). The X-ray dot map in Figure 5-2 of the same field of view illustrates the utility of this technique in accurately delineating the region of metal concentration. Figures 5-1 and 5-2 show that the waste inclusions are in the micron range and perhaps smaller and that the cadmium and chromium sometimes concentrate separately, possibly indicating a chemical rather than a physical mechanism for the metal concentration.

Another useful technique is the X-ray line scan which can indicate elemental concentration differences along a specified line on the sample surface. This has particular applications in microscopic characterization of leaching behavior. It is possible, in principle, to study concentration gradients of a waste metal from the leached cement surface to the unleached interior.

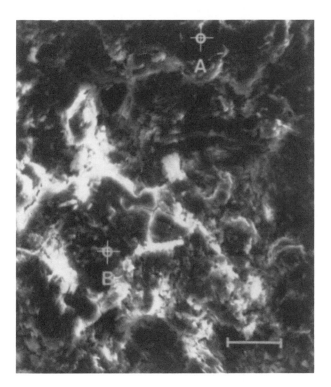

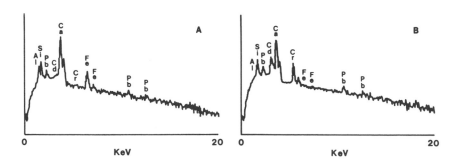

Figure 5-1.   SEM micrographs of cement paste (secondary electron or SE image) and X-ray spectra of the points indicated. 35 keV, 0.2 nA, 2000-s acquisition time, log intensity scale, bar = 30 μm.

Figure 5-3 shows an electron micrograph of a fractured cement piece, containing Cr, which has been leached by the toxicity characteristics leaching procedure (TCLP). The accompanying X-ray spectra clearly show that the outer surface has a lower Cr concentration relative to the unleached interior

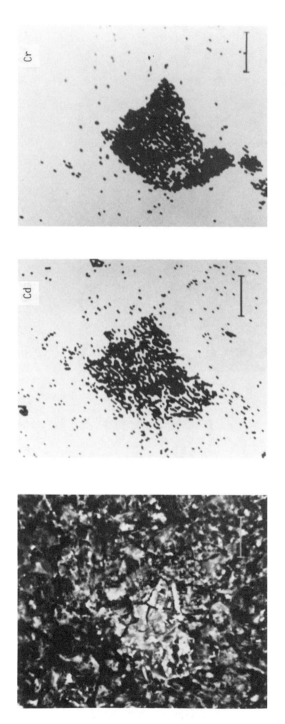

Figure 5-2.   SEM micrograph of cement paste (BSE image), the same field of view as Figure 5-1. The accompaning dot maps show Cd and Cr concentrations.

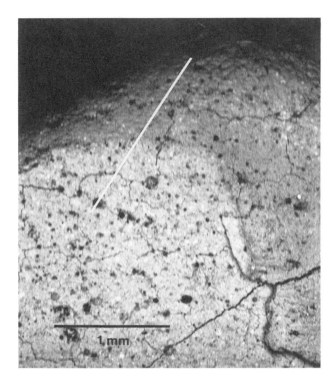

Figure 5-3.   Backscattered electron image of a leached, fractured cement particle. The fracture has exposed the relatively higher
Cr content interior, which appears brighter due to a higher
average atomic number than the leached exterior. The
linescan region corresponding to Figure 5-5 is shown.

(Figure 5-4). Figure 5-5 shows an X-ray line scan for Fe and Cr across the
leached and unleached surface. This line scan (across the sample as shown in
Figure 5-3) shows a drop in Cr concentration at the boundary. The fact that Fe
concentration is relatively constant indicates that the change in Cr content is
due to leaching and not to the geometry of the fracture surface and that leaching
effects from the TCLP are still observable after sample attrition.

Transmission and Scanning Transmission Electron Microscopy (TEM/STEM)

TEM Specimen Preparation

A relatively simple and reproducible technique for the preparation of thin
foil specimens from cementitious materials has been developed. This method
involves the use of a combined dimpling/ion milling process which permits

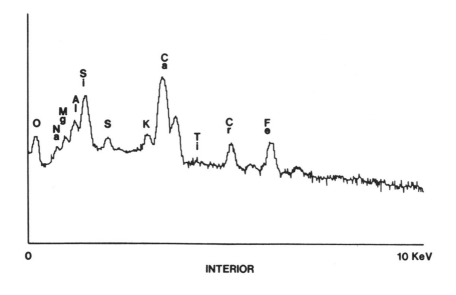

INTERIOR

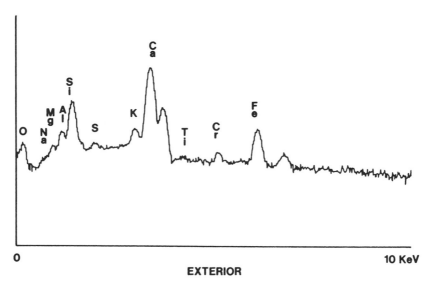

EXTERIOR

Figure 5-4.   X-ray spectra (displayed on a log scale) of the interior and exterior areas of the leached cement shown in Figure 5-3, which illustrate the significant decrease in Cr content at the cement surface due to the leaching process. The spectra were acquired at 25 kV and 0.2 nA for 200 s.

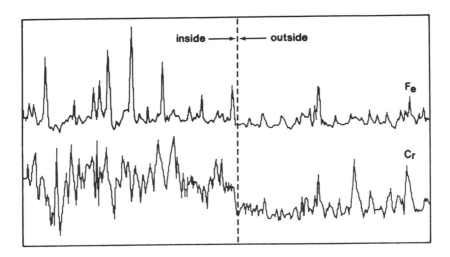

Figure 5-5.   X-ray linescan across the interior/exterior fracture boundary of Figure 5-3, showing the distinct decrease in Cr on
the exterior surface of the leached cement. The linescan
was acquired over 256 points, corresponding to the line
(1600 mm) in Figure 5-3. The X-ray intensities were determined from the K_α lines of Cr and Fe, acquired at 25 kV,
0.2 nA, and 20s per point.

thinning of the center of the specimen while maintaining a relatively thick outer
rim, thereby facilitating specimen handling. Specimens, prepared by this technique, are thin enough to be examined in a 100-kV electron microscope,
eliminating the need for the somewhat more expensive 1000-kV instrument.

The specimen preparation process is detailed in References 11 to 13. Minor
improvements have been made subsequent to these publications, and these are
included in the following description. The various preparation steps are also
depicted in the SEM micrographs of Figure 5-6.

1.   A 1- to 2-mm thick slice is cut from the cured material using a low speed
       diamond saw, followed by cleaning in ethanol.
2.   3-mm discs are machined from this slice, using a Gatan Model 601
       ultrasonic disc cutter (Figure 5-6a). The cutting action is provided by a
       SiC slurry located between the specimen surface and the aluminum drill
       bit.
3.   The discs are mechanically polished, from both sides, to a thickness of
       150 to 250 μm (Figure 5-6b). Initially 30-μm SiC lapping discs are used
       to remove most of the material, while final polishing is done on 5-μm
       SiC discs. Care is taken to ensure that the surfaces are parallel.

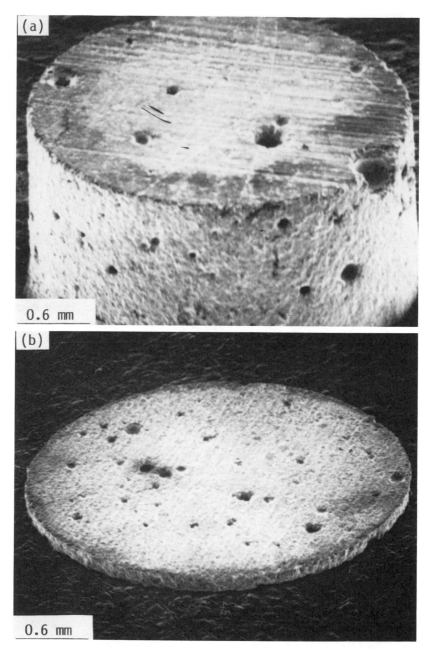

Figure 5-6.   SEM micrographs depicting the sequence of TEM specimen preparation. See text for details. (From Ivey, D. G.,
R. B. Heimann, M. Neuwirth, S. Shumborski, D. Conrad,
R. J. Mikula, and W. W. Lam. 1990. *J. Mater. Sci.* 25:
5055–5062. With permission.)

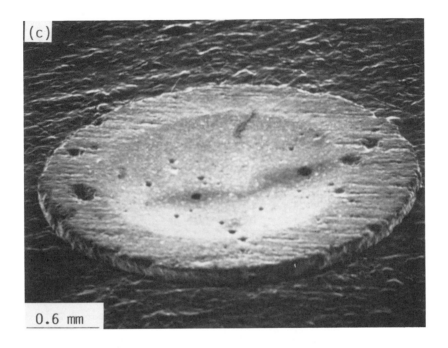

Figure 5-6 continued.

4.   Specimens are dimpled, on one side, down to a central thickness of about 70 µm, using 0.25-µm diamond paste (Figure 5-6c). The dimpler used here is a Gatan Model 656/3 Dimple Grinder, although other comparable brands are also available. Prior to dimpling, the specimen is mounted on a stainless steel plug with a low melting point wax that dissolves in acetone. The mounted specimen is subsequently placed on a magnetic turntable, which allows the specimen to rotate within its own plane. A 15-mm diameter phosphor bronze polishing wheel with spherical edges is coated with a small amount of 0.25-µm diamond paste and lowered onto the surface of the specimen. Using a low rotation speed (2 to 3 on a scale of 10) and a small load (20 g), the polishing wheel gently machines a spherical depression in the center of the disc.

5.   The other side is then dimpled in the same manner until a small hole appears (<0.25 mm in diameter). Because of the large radius of curvature of the polishing wheel, the region immediately surrounding the hole is fairly flat and optically transparent.

6.   Specimens are removed from the plug, thoroughly rinsed in acetone to dissolve the wax, followed by rinsing in ethanol. Each disc is then bonded to copper rings (on either side), for durability and support, with a two part, rapid setting epoxy (Figure 5-6d).

7.   Specimens are ion milled for 1.5 to 2 h, using Ar as the sputtering species, at a potential of ≈4.5 to 5.0 kV, a current of ≈0.5 mA per gun, and an incidence angle of 18° to the horizontal. Specimen cooling during ion milling (liquid nitrogen) helps reduce heating effects during the sputtering process. Ion milling may produce radiation damage effects, such as Ar implantation and defect generation. Small amounts of Ar have been detected in many of our samples. These effects can be minimized by reducing the accelerating voltage and gun current during the final stages of preparation (3 kV and 0.2 mA per gun). The total preparation time for one specimen is between 4 and 6 h.

8.   Ion-milled specimens are then coated with a thin layer of evaporated carbon to reduce charging effects during TEM examination. The final specimens contain electron transparent regions at operating voltages as low as 100 kV.

TEM/STEM Examination

The TEM used in most of our work was a Hitachi H-600 TEM/STEM, which is equipped with a Be window X-ray detector. The maximum accelerating voltage is 100 kV, and this setting was used at all times to provide maximum electron penetration.

A brief description of the capabilities of a typical TEM/STEM is given below. More detailed descriptions are available in standard TEM texts (e.g.,

References 14 and 15). A TEM/STEM combines many of the imaging and analytical capabilities of an SEM with the imaging and diffraction capabilities of a conventional TEM.

Quantification of X-ray data is often more straightforward in a TEM/STEM compared to an SEM, because the specimen is so thin. For a thin foil specimen, absorption and fluorescence effects can be neglected so that the ratio of concentrations of two elements is simply proportional to the ratio of their X-ray intensities:[16]

$$\frac{C_A}{C_B} = k_{A/B} \frac{I_A}{I_B}$$

C_A and C_B are the concentrations in weight percent of elements A and B, I_A and I_B are the respective X-ray intensities, and $k_{A/B}$ is the proportionality or Cliff-Lorimer factor. The "k" factor is a function of the specimen and microscope operating conditions, and can be determined from a standard specimen of known composition. The spatial resolution is much better than in the SEM, due to the thin nature of the specimen. The resolution is ultimately dependent on beam (e.g., size, accelerating voltage) and specimen (e.g., type, thickness) characteristics. Elements with atomic numbers as low as $z = 11$ (Na) are detectable with a conventional Be window X-ray detector. Light element detection is possible (down to B) with the newer windowless or ultra-thin window detectors.

An additional analytical technique is available on several microscopes, i.e., electron energy loss spectroscopy or EELS. An EELS spectrometer, which is located below the specimen, analyzes the energy lost by the transmitted electrons. The energy loss is characteristic of a particular electron in a particular atom. EELS is particularly useful for light element detection, although the data are usually more difficult to interpret and quantify.

When operating in TEM mode, image contrast can arise in three ways, i.e., absorption contrast, diffraction contrast, and phase contrast. Absorption contrast is due to scattering (both inelastic and elastic) of electrons as they pass through the specimen. Contrast differences arise from sample thickness, atomic number, and density variations in the specimens. Image resolution is the poorest of the three techniques. Diffraction contrast arises from Bragg diffraction of the electron beam as it passes through the specimen. To obtain a bright field image an aperture is placed in the back focal plane of the objective lens, allowing only those electrons that pass through the specimen without being diffracted to contribute to the image. Regions that strongly diffract electrons show up dark, while regions that do not scatter electrons appreciably show up bright. The best resolution of this technique is about 1.5 nm, making it particularly useful for characterizing crystal defects, such as grain boundaries, phase boundaries, and dislocations. Diffraction contrast is the most commonly used

TEM imaging technique for crystalline materials. Phase contrast images are due to interference effects between recombined unscattered and diffracted electron beams and provide the best resolution (0.2 to 0.3 nm — atomic dimensions). These images, however, are not routinely attainable in most multipurpose analytical electron microscopes, due in part to specimen thickness limitations.

Crystallographic information from individual grains or precipitates is attainable by means of electron diffraction. The principles are very similar to those for X-ray diffraction, although Bragg angles are considerably smaller (<1°). Two diffraction techniques are available on modern microscopes, i.e., selected area diffraction (SAD) and convergent beam electron diffraction (CBED). In SAD, the region of interest is defined by an appropriately sized aperture placed in the image plane of the microscope. The resultant pattern consists of an array of spots, with each spot corresponding to diffraction from a particular crystallographic plane. The best attainable resolution is $\cong 0.5$ μm. In CBED, the region of interest is defined by the size of the electron probe. The electron beam is focused onto the specimen surface and, as such, the resolution depends on the probe size. Resolution also depends on other factors such as specimen thickness and atomic number — an increase in thickness or atomic number leads to beam spreading — but is often better than 20 nm. CBED has the additional benefit of providing three-dimensional crystallographic information. CBED patterns are similar to SAD patterns in appearance, except the spots are broadened into discs due to convergence of the electron beam.

Typical electron transparent regions from an ion milled specimen of hydrated OPC are shown in Figures 5-7 and 5-8. A region of unreacted alite (impure C_3S^*) is visible in Figure 5-7. Alite is a major constituent of OPC clinker, making up about 55% by weight of the clinker material. The composition was determined by X-ray microanalysis, while the crystal structure was determined from SAD patterns obtained at different orientations (Figure 5-7). Alite, in all cases, appeared as the monoclinic M1-type structure, with lattice parameters; a = 3.308 nm, b = 0.707 nm, c = 1.856 nm, and β = 94.17°. The intense reflections in the selected area diffraction pattern (SAD) of Figure 5-7 arise from the Jeffrey rhombohedral pseudostructure, with hexagonal axes of a = 0.7 nm and c = 2.50 nm. The less-intense reflections are from the monoclinic structure. Figure 5-8 depicts a region containing free lime (CaO) and calcite ($CaCO_3$). These regions were also identified by X-ray microanalysis and electron diffraction.

5.4 PRELIMINARY RESULTS ON CR³⁺ CONTAINMENT

Most of our work to the present time has concentrated on Cr^{3+} containment in OPC.[12,13] Chromium was chosen for several reasons. Many industrial liquid

---

* Cement chemistry notation: C = CaO, S = SiO_2, A = Al_2O_3, F = Fe_2O_3, H = H_2O,
  $(\bar{S})$ = SO_3.

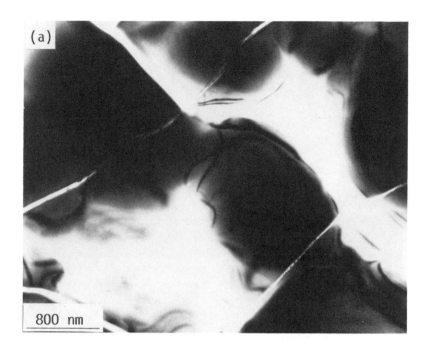

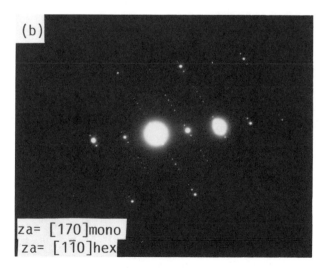

Figure 5-7.   (a) Bright field micrograph and (b) SAD pattern of a region of unreacted alite in 1.0 *M* Cr cement paste. The beam directions for both the monoclinic and hexagonal cells are given. (From Ivey, D. G., R. B. Heimann, M. Neuwirth, S. Shumborski, D. Conrad, R. J. Mikula, and W. W. Lam. 1990. *J. Mater. Sci.* 25: 5055–5062. With permission.)

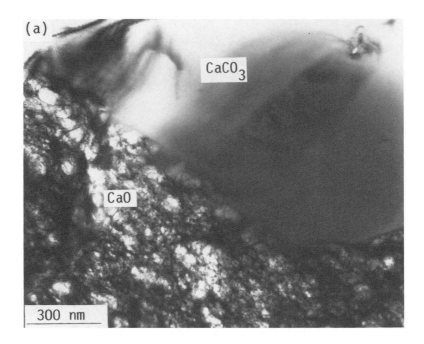

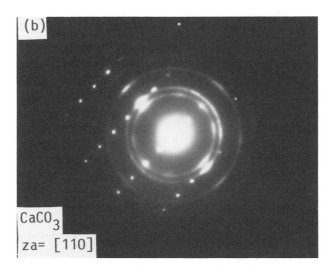

Figure 5-8.   (a) Bright field image and (b) SAD pattern of region of calcia that has partially reacted to form calcite. The spot pattern in (b) is from single crystal calcite, while the ring pattern is from the fcc calcia phase. (From Ivey, D. G., R. B. Heimann, M. Neuwirth, S. Shumborski, D. Conrad, R. J. Mikula, and W. W. Lam. 1990. *J. Mater. Sci.* 25: 5055–5062. With permission.)

waste streams contain Cr, which is potentially toxic and environmentally incompatible. It is generally only a very low-level contaminant in Portland cement (<50 ppm), and it has a characteristic K_α X-ray peak which is sufficiently isolated from any X-ray peaks generated by other cement constituents. Both of these features facilitate identification of the metal species. Also, Cr is known to accelerate the hydration reaction and strength development in cement,[17,18] which may be of practical significance.

Samples of OPC were mixed with 0.1, 0.4, and 1.0-M solutions of $Cr(NO_3)_3 \cdot 9H_2O$, at a solution/cement ratio of 0.33. A control sample, i.e., OPC mixed with distilled water with the same solution/cement ratio was also fabricated. The samples were mixed and cast in cylindrical molds, 45 mm inside diameter by 75 mm high. Samples (500 to 600 g in weight) were cured at 96% relative humidity for 36 d at 25°C.

No concentration effects were detected in waste-containing cement specimens, other than in the frequency of detection of Cr.

SEM analysis indicated differences in the gross microstructure between the control and Cr-containing specimens. The major difference between the two types of specimens was the absence of ettringite or any of the calcium sulfoaluminate related phases in the Cr-containing cement paste (Figure 5-9). Chromium appeared to inhibit the formation of ettringite, which may be related to its role as a hydration accelerator. Gypsum, a known retarding additive, is believed to promote ettringite formation.[2] Ettringite hinders the formation of an alumina-silica gel, which is responsible for the initial rapid set. Chromium, being an accelerator, may operate in the opposite way. Note, also, in Figure 5-9 the presence of the two major products of hydration, i.e., calcium-silicate-hydrate (C-S-H) and calcium hydroxide (CH).

Chromium was detected virtually everywhere in all the SEM samples. The levels of Cr were quite variable as indicated in the BSE image and accompanying X-ray spectra (Figure 5-10). There appears to be a correlation between the Si and Cr concentrations in the cement paste. As the Cr content increases, the Si content decreases, indicating that Cr may have substituted for Si. Further evidence for this was found in the TEM/STEM analysis, which will be discussed in the following paragraphs. Individual phase compositions were difficult to determine because of marked electron beam spreading in the bulk SEM specimens. This problem was not a factor in the TEM/STEM analysis due to the thin nature of the specimens.

Analysis in the TEM/STEM provided a much more detailed, although less statistically representative, view of the hydrated cement paste. All of the major hydration products, with the exception of ettringite (AFt phases), were observed in both the control and Cr-containing cement samples. Chromium was detected in virtually all the hydration products, although not in all instances. The amount of Cr, relative to Ca and Si, was quite variable, echoing the SEM results.

C-S-H and CH are the most abundant hydration products and these were the most frequently detected phases in the control and Cr-containing specimens.

Figure 5-9.   SEM micrographs of cement paste specimens: (a) control specimen and (b) 1.0 M Cr specimen. Note, the presence of ettringite needles in (a) and the absence of ettringite in (b). (From Ivey, D. G., R. B. Heimann, M. Neuwirth, S. Shumborski, D. Conrad, R. J. Mikula, and W. W. Lam. 1990. *J. Mater. Sci.* 25: 5055–5062. With permission.)

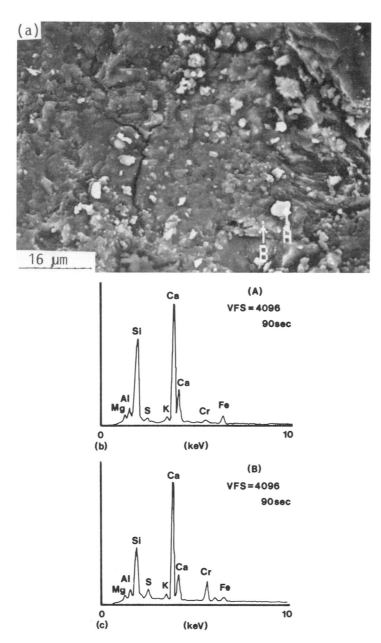

Figure 5-10.   (a) BSE image of a region of 1.0-*M* Cr cement paste. (b) X-ray spectra corresponding to regions marked A and B in (a). Note, the apparent inverse relationship between Si and Cr. (From Ivey, D. G., R. B. Heimann, M. Neuwirth, S. Shumborski, D. Conrad, R. J. Mikula, and W. W. Lam. 1990. *J. Mater. Sci.* 25: 5055–5062. With permission.)

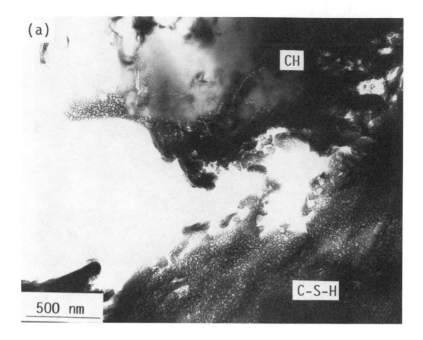

Figure 5-11.   (a) Bright field micrograph, (b and c) SAD patterns, and (d and e) X-ray spectra of a 1.0-*M* Cr cement paste, showing areas of C-S-H and CH. Both phases contain Cr. (From Ivey, D. G., R. B. Heimann, M. Neuwirth, S. Shumborski, D. Conrad, R. J. Mikula, and W. W. Lam. 1990. *J. Mater. Sci.* 25: 5055–5062. With permission.)

Calcite was also prevalent, forming from the reaction of unprotected CH or CaO in the presence of air.

A region of C-S-H and CH is shown in Figure 5-11, along with accompanying SAD patterns. Both phases contained Cr, although the CH had a significantly higher Cr concentration. Other regions of C-S-H, with much higher levels of Cr, were also observed. An example is shown in Figure 5-12, where the accompanying X-ray spectra also demonstrate the wide variability in phase composition. There appeared to be a correlation between the Cr and Si concentrations, as mentioned previously in the SEM results. A decrease in Si content appeared to be associated with a corresponding increase in Cr content.

A possible explanation for the above observations would be the substitution of Cr (trivalent) for Si (tetravalent) in the silicate tetrahedra. Fe^{3+} and Al^{3+} are also known to substitute for Si in C-S-H. The charge imbalance may be offset by the presence of other ions. For example, a H^+ ion or a hydrated proton H_3O^+ may be associated with each substituted Cr^{3+}. Other ions such as Mg^{2+} may be involved, in a more complex manner, as well. The ionic radius ratio of Cr/O

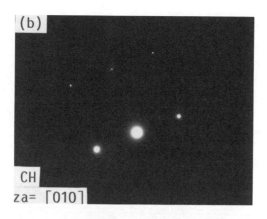

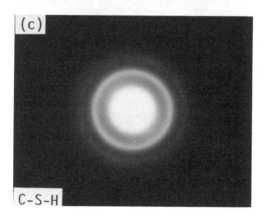

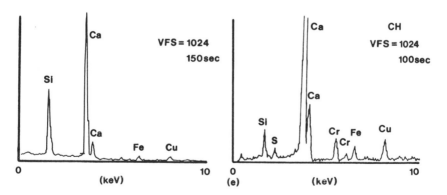

Figure 5-11 continued.

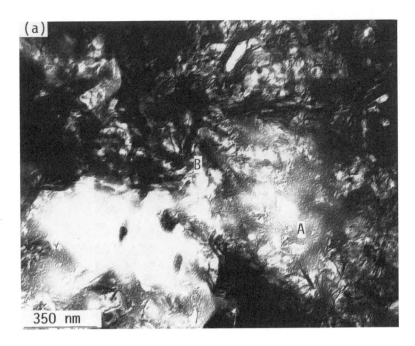

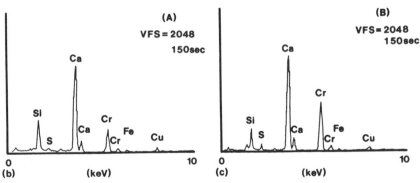

Figure 5-12.   (a) Bright field micrograph and (b) X-ray spectra of a high Cr region of C-S-H in a 1.0-*M* Cr cement paste. Note, from the X-ray spectra in (b), that Cr appears to substitute for Si. (From Ivey, D. G., R. B. Heimann, M. Neuwirth, S. Shumborski, D. Conrad, R. J. Mikula, and W. W. Lam. 1990. *J. Mater. Sci.* 25: 5055–5062. With permission.)

is favorable for either tetrahedral or octahedral coordination according to Pauling's rules,[19] which provides further evidence that chromate ion formation is indeed possible. Cr^{3+} substitution into C-S-H has been reported previously.[20,21] Tashiro et al.[20,21] have reported that 10% or more of Cr_2O_3 (Cr^{3+})

can enter into solid solution with C-S-H. They, however, suggested that Cr^{3+} substitutes for both Ca^{2+} and Si^{4+}, i.e., $2Cr^{3+} = Ca^{4+} + Si^{4+}$. We do not discount this possibility, as our results at this time are somewhat limited. A much more detailed analysis of Cr^{3+} substitution in C-S-H is necessary for clarification.

Hydration products of C_3A and C_4AF have also been detected by TEM/STEM, although much more infrequently than C-S-H or CH. In agreement with the earlier SEM results, only the monosulfates (AFm) and not the trisulfates (AFt) were observed. Significant amounts of Cr were detected in these phases. Chromium likely substituted for both Al and Fe, since the ionic charges are the same (3+) and the ionic sizes are comparable.

Additional phases, not found in the hydrated control samples, were present in the waste samples. One of these phases was polycrystalline (as indicated by the SAD pattern in Figure 5-13) and Ca-Cr rich (Figure 5-13). This phase was fairly abundant, i.e., more so than the monosulfate phases. Once again, the composition was quite variable, with a maximum Cr/Ca ratio of $\cong 0.85$ (X-ray spectra in Figure 5-13). Other elements present in this phase included Si, Fe, and S. We have been unable, at this time, to identify this phase, mainly because of the wide compositional variability.

The other phase, shown in Figure 5-14, was only observed in one sample. It was single crystal, as indicated in the convergent beam electron diffraction (CBED) pattern in Figure 5-14. The diffraction pattern matched with the crystal structure of $Ca(NO_3)_2$. Chromium, as well as other impurities such as Si, Fe, and S, were also present. The source of the nitrate ions was likely the original chromium solution.

From the preceding results, it appears that Cr^{3+} is chemically, rather than physically, contained within the cement structure. This appears to be a feature of accelerating additives, whereas retarders tend not to be chemically contained. Lead (Pb^{2+}), for example, is a known hydration retarder and has been found, by photoelectron spectroscopy, to be primarily a surface species in cement paste.[23] This is in contrast to Cr, which in the same study was barely detectable at the surface. Also, Zn greatly retards hydration of OPC (and pulverized fuel ash, PFA) by forming a layer of amorphous gel on particles with concomitant promotion of ettringite formation that results in a drastic increase in pore volume and the proportion of pores with large radii.[23] Retarders form insoluble salts and tend to precipitate as dense coatings on the hydrating phases.[17,22] This results in a loss of permeability, effectively forming a diffusion barrier to water. The net result is to slow the rate of hydration. Accelerators, such as Cr, have soluble calcium salts and thus are more easily incorporated into the hydration products.

5.5 SUMMARY

The capabilities and the complementary nature of SEM and STEM in characterizing cementitious materials have been demonstrated. Specifically,

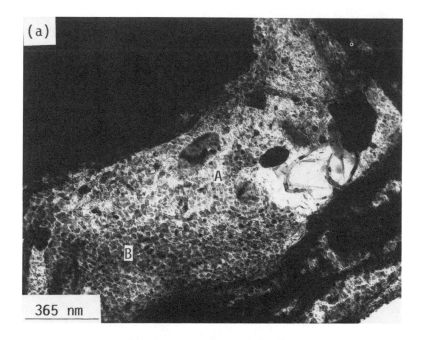

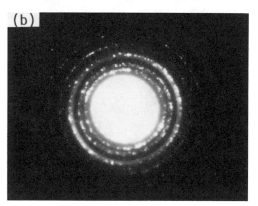

Figure 5-13.   (a) Bright field image, (b) SAD pattern, and (c) X-ray spectra from a polycrystalline Ca-Cr rich region. (From Ivey, D. G., R. B. Heimann, M. Neuwirth, S. Shumborski, D. Conrad, R. J. Mikula, and W. W. Lam. 1990. *J. Mater. Sci.* 25: 5055–5062. With permission.)

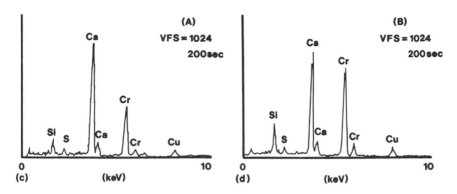

Figure 5-13 continued.

these techniques are shown to be useful in probing the distribution and mechanisms of containment of metal wastes in various cement phases.

Preliminary studies of chromium (trivalent) containment in OPC indicate that chromium is dissolved in all of the major hydration products. In the case of calcium-silicate-hydrate, chromium appears to substitute for silicon in the silicate complexes. Chromium also is found to concentrate in a Ca-Cr rich polycrystalline phase.

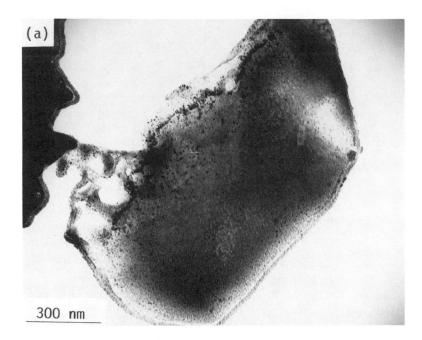

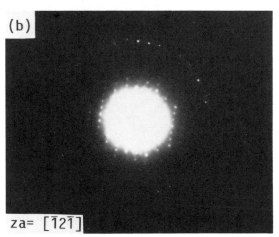

Figure 5-14.   (a) Bright field image, (b) CBED pattern, and (c) X-ray spectrum of a region of Cr-containing Ca(NO₃)₂. (From Ivey, D. G., R. B. Heimann, M. Neuwirth, S. Shumborski, D. Conrad, R. J. Mikula, and W. W. Lam. 1990. *J. Mater. Sci.* 25: 5055–5062. With permission.)

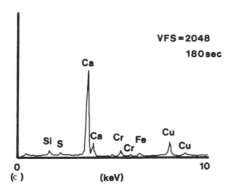

Figure 5-14 continued.

REFERENCES

1. Lachowski, E. E. and S. Diamond. 1985. "Investigation of the Composition and Morphology of Individual Particles of Portland Cement Pastes." *Cem. Concr. Res.* 13: 177–185.
2. Eaton, H. C., M. B. Walsh, M. E. Tittlebaum, F. K. Cartledge, and D. Chalasani. 1985. "Microscopic Characterization of Solidification/Stabilization of Organic Hazardous Wastes." Paper presented at Energy Sources and Technology Conference and Exhibition, Dallas, Texas, February 17–21 (Publication of ASME 85-Pet-4).
3. Lachowski, E. E., K. Mohan, H. F. W. Taylor, and A. E. Moore. 1980. "Analytical Electron Microscopy of Cement Pastes. II. Pastes of Portland Cements and Clinkers." *J. Am. Cer. Soc.* 63(7-8): 447–452.
4. Barnes, P. and A. Ghose. 1983. "The Microscopy of Unhydrated Portland Cements" in *Structure and Performance of Cements.* Barnes, P., Ed. Applied Science, New York. pp. 139–203.
5. Sinclair, W. and G. W. Groves. 1984. "Transmission Electron Microscopy and X-Ray Diffraction of Doped Tricalcium Silicate." *J. Am. Cer. Soc.* 67(5): 325–330.
6. Lawrence, F. V., Jr., D. A. Reid, and A. A. de Carvalho. 1974. "Transmission Electron Microscopy of Hydrated Dicalcium Silicate Thin Films." *J. Am. Cer. Soc.* 57(3): 144–148.
7. Jennings, H. M., B. J. Dalgleish, and P. L. Pratt. 1981. "Morphological Development of Hydrating Tricalcium Silicate as Examined by Electron Microscopy Techniques." *J. Am. Cer. Soc.* 64(10): 567–572.
8. Double, D. D. 1973. "Some Studies of the Hydration of Portland Cement Using High Voltage (1 MV) Electron Microscopy." *Mater. Sci. and Eng.* 12: 29–34.

9.   Groves, G. W., P. J. Le Sueur, and W. Sinclair. 1986. "Transmission Electron Microscopy and Microanalytical Studies of Ion-Beamed-Thinned Sections of Tricalcium Silicate Pastes." *J. Am. Cer. Soc.* 69(4): 353–356.

10. Neuwirth, M., R. J. Mikula, and P. Hannak. 1989. *Comparitive Studies of Metal Containment in Solidified Matrices by Scanning and Transmission Electron Microscopy.* Gilliam, T. M. and P. Cote, Eds. American Society for Testing and Materials, Philadelphia. pp. 201–213.

11. Ivey, D. G. and M. Neuwirth. 1989. "A Technique for Preparing TEM Specimens From Cementitious Materials." *Cem. Concr. Res.* 19: 642–648.

12. Ivey, D. G., R. B. Heimann, M. Neuwirth, S. Shumborski, D. Conrad, R. J. Mikula, and W. W. Lam. 1990. "Electron Microscopy of Heavy Metal Waste in Cement Matrices." *J. Mater. Sci.* 25: 5055–5062.

13. Ivey, D. G., S. A. Shumborski, and M. Neuwirth. 1989. "TEM/STEM Analysis of Cement Stabilized Heavy Metal Waste." Paper presented at Scanning 89/EM West Meeting, Long Beach, CA, April 5–7.

14. Von Heimendahl, M. 1980. *Electron Microscopy of Materials, An Introduction.* Academic Press, New York.

15. Edington, J. W. 1976. *Practical Electron Microscopy in Materials Science.* Van Nostrand Reinhold, New York.

16. Goldstein, J. I., D. B. Williams, and G. Cliff. 1986. *Quantitative X-Ray Analysis, from Principles of Analytical Electron Microscopy.* Joy, D. C., A. D. Romig, and J. I. Goldstein, Eds. Plenum Press, New York.

17. Thomas, N. L. 1987. "Corrosion Problems in Reinforced Concrete: Why Accelerators of Cement Hydration Usually Promote Corrosion of Steel." *J. Mater. Sci.* 22: 3328–3334.

18. Young, J. F., Ed. 1982. *Cements Research Progress, 1981.* American Ceramic Society, Columbus, OH. pp. 60–61.

19. Kingery, W. D. 1960. *Introduction to Ceramics.* John Wiley & Sons, New York. pp. 109–113.

20. Tashiro, C. and K. Kawaguchi. 1977. "Effects of CaO/SiO_2 Ratio and Cr_2O_3 on the Hydrothermal Synthesis of Xonotlite." *Cem. Concr. Res.* 7: 69–76.

21. Tashiro, C., H. Takahashi, M. Kanaya, I. Hirakida, and R. Yoshida. 1977. "Hardening Property of Cement Mortar Adding Heavy Metal Compounds and Solubility of Heavy Metal from Hardened Mortar." *Cem. Concr. Res.* 7: 283–290.

22. Cocke, D., J. D. Ortego, H. McWhinney, K. Lee, and S. Shukla. 1989. "A Model for Lead Retardation of Cement Setting." *Cem. Concr. Res.* 19: 156–159.

23. Poon, C. S. and R. Perry. 1987. "Studies of Zinc, Cadmium and Mercury Stabilization in OPC-PFA Mixtures." *Mat. Res. Soc. Symp. Proc.* 86: 67–76.

6 MICROSCOPIC AND NMR SPECTROSCOPIC CHARACTERIZATION OF CEMENT-SOLIDIFIED HAZARDOUS WASTES

L. G. Butler, F. K. Cartledge, H. C. Eaton, and M. E. Tittlebaum

6.1. ABSTRACT

A number of techniques give useful chemical information about solidified wastes, including in particular scanning electron microscopy, X-ray diffraction, infrared spectroscopy, and solid-state nuclear magnetic resonance (NMR) spectroscopy. Solidified samples prepared using cements and various additives, and including both inorganic and organic materials as wastes, have been examined and have yielded information about the chemical speciation of the wastes in the matrix and about the effects which the waste additives have upon the cement matrix itself. These kinds of information are potentially very important for both understanding and improving the performance of solidification/stabilization (S/S) technology. Specific examples span a range of possibilities from (1) the case of a mixed metal sludge where solidification in cement results in encapsulation only and little evidence of chemical interaction between waste and matrix, to (2) cases such as Cr(III) and phenols where there are clear changes in the cement matrix and also in the chemical forms of the wastes within the matrix.

0-87371-748-1/93/$0.00+.50
© 1993 by Lewis Publishers

6.2. INTRODUCTION

S/S of hazardous wastes is a widely used technology, and it is clearly important to be able to evaluate its effectiveness and to make rational attempts to improve its performance. For a number of years, we have been conducting research that has as a principal aim the understanding of the detailed chemistry of S/S systems. In practical terms, the two most important questions to address are the extent of immobilization of hazardous constituents and the permanence of the matrix structure and properties under environmental stresses. Clearly, one would ideally like to be able to relate the former (the chemistry of the system) to the latter (practical performance), and that has been a major goal of our research.

In practice, of course, because of the enormous complexity of S/S systems it is not possible to characterize fully either the chemistry or the performance. Consequently, we have posed more limited questions, such as the following:

1. Since S/S involves containment of waste constituents in a matrix, can we characterize the matrix as fully as possible, and can we define effects which the waste admixtures have on the structure of the matrix?
2. What chemistry, if any, do the waste constituents undergo during S/S?
3. Can we define where in the matrix the wastes are found and with what degree of homogeneity?
4. Are there any clear correlations between the answers to the questions above and measures of performance, such as leachability and strength?

The questions above are already too broad to get completely satisfactory answers to, but the descriptions below will give an indication of what we are capable of doing at present and how we are continuing to pursue these thorny problems.

The binding agents in use in S/S vary from completely inorganic, cement-like materials to thermoplastic organics, and a variety of additives are also common.[2] Thus, a major complication to detailed study of the process (indeed, a complication to the practical applications as well) is the complexity of the mixtures involved, and in some cases also the variability from batch to batch of major ingredients such as fly ash and kiln dust. In the current work we have restricted ourselves mostly to relatively simple Portland cement systems with a few common additives.

However, even Portland cement alone is already a formidably complex mixture that defies complete characterization. Any specific technique gives only limited information. Consequently, we have used a variety of microscopic and spectroscopic methods to generate complementary sets of information. Combining these sets into a consistent, meaningful interpretation of the behavior of the whole system is a challenging process, as the examples below will illustrate.

6.3. CEMENTS WITH METALLIC WASTES

Portland cement (with or without additives) combined with a single metal salt has often been used as the simplest approximation to a S/S system in a great deal of exploratory work. For instance, we have investigated the combination of a number of soluble heavy metal salts, as well as sludges formed by treatment of the soluble salts with lime, with type I Portland cement (OPC). The resulting solid products have been characterized using a number of physical techniques, including compressive strength testing and both static and dynamic leaching procedures. Techniques familiar to cement chemists include scanning electron microscopy (SEM), conduction calorimetry, and infrared spectroscopy, and we have used them along with more modern advances, such as solid-state NMR spectroscopy. The physical tests are crude measures of performance. SEM can be used to follow the formation of the major phases characteristic of hydrated cement paste and identifiable from their morphologies. Calorimetry has often been used to follow cement hydration reactions, since peaks in the plots of heat output vs. time can be correlated with changes in physical properties and the appearance of particular phases of the mature cement structure. The latter can often be seen in the SEM.

Solid-state NMR spectroscopy employing magic angle spinning has been used in our studies in several different ways. The solid-state ^{29}Si and ^{27}Al MAS NMR spectra contain peaks, often overlapping ones, that correspond to Si in five different environments and Al in two different environments. The ^{29}Si chemical shifts depend mainly on the degree of condensation of silicon-oxygen tetrahedra.[3-5] The symbol Q represents a Si atom surrounded by four oxygen atoms. A superscript following the Q shows the number of other Q units attached to the Si atom under study. The principal transformation occurring during cement hydration, as followed by NMR, is the formation of chains containing Q^1 (chain-terminating) and Q^2 (chain lengthening) units starting from the orthosilicate ions (Q^0) present in the cement clinker.[6,7] The loss of Q^0 units as a function of time is a measure of the overall degree of hydration of the cement clinker and agrees well with calorimetric measurements of degree of hydration.[7] Mature cement paste without metal additives contains mainly Q^1 units, with a much smaller proportion of Q^2 units, and little or no Q^3 (chain branching) or Q^4 (crosslinking) units.

Hydration of aluminate phases, which account for roughly 10% of the cement clinker, occurs more rapidly than that of the silicate phases. The chemical shift differences between the 4-coordinate Al atoms present in the clinker and the 6-coordinate Al atoms in the hydrated aluminates can be followed as a function of time in the ^{27}Al NMR spectra, as was initially shown for hydration of tricalcium aluminate.[8] In all samples of cement clinker that we have examined, both 4- and 6-coordinate Al are present initially, and different lots from different manufacturers show wide variation in the ratio of the two types of Al peaks. Hydration almost always converts the mixture essentially completely to 6-coordinate Al.

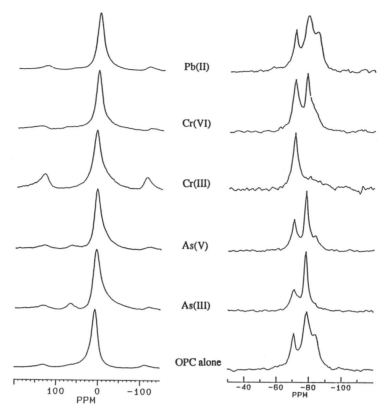

Figure 6-1. ^{27}Al (left) and ^{29}Si (right) solid-state MAS NMR spectra
obtained at 4.7 T and spin rates of 8 (Al) and 5 (Si) KHz for
type I Portland cement samples cured for 28 d and containing 10% by weight of As (III), As (V), Cr (III), Cr (VI), and
Pb (II).

Chemical shift information is also useful for identifying the chemical form
of the waste in the cement matrix. One example of this kind of study will be
described below with phenols in cement, but there are many other possibilities.

A second, qualitatively different, kind of NMR information comes from
proton relaxation time measurements, which can define the number of different
environments for hydrogen atoms (contained in water molecules or OH$^-$ ions)
that are present in the cement paste, and the proportions of the total in each
environment. These techniques have been applied to cement hydration previously.[9-11]

One series of samples which illustrates our ability to characterize the cement
matrix by NMR is shown in Figure 6-1. The samples are all aged for 28 d, have
the same water/cement ratio (0.5), and contain 10% by weight of a heavy metal
as an admixture. The specific salts added are $NaAsO_2$, $Na_2HAsO_4 \cdot 7H_2O$,

$Cr(NO_3)_3 \cdot 9H_2O$, $Na_2CrO_4 \cdot 4H_2O$, and $Pb(NO_3)_2$. All NMR spectra were obtained with a Bruker MSL-200 spectrometer operating at 4.7 Tesla (200 MHz). The ^{27}Al spectra were obtained using approximately 150 mg of powdered sample in 7.5-mm Torlon rotors spinning at 8 KHz, and chemical shifts are reported in ppm relative to external aq. $AlCl_3$. The ^{29}Si spectra were obtained using approximately 300 mg of sample in 7-mm zirconia rotors spinning at 5 KHz, and are reported relative to external tetramethylsilane. From many determinations using duplicate and triplicate samples, integrations under the peaks are reproducible to ±4%.

The ^{27}Al spectra are shown on the left in Figure 6-1. In these spectra after 28 d of cure, all the spectra are dominated by the octahedral Al peak (Al[6]) centered at about 5 ppm (and usually showing symmetrically disposed spinning side bands). The As(III) sample clearly shows the remains of the tetrahedral Al peak (Al[4]) centered at about 60 ppm. Thus, the Al spectra do not reveal very significant differences among the matrices. However, the ^{29}Si spectra, on the right in Figure 6-1, are profoundly different, and the silicate species giving rise to the spectra constitute the major fraction of the cement matrix. All the spectra show some evidence of three silicate species with chemical shifts centered at −70, −79, and −84 ppm, corresponding to Q^0, Q^1, and Q^2, respectively. The visual impression of the relative proportions of the various silicate units that can be obtained from Figure 6-1 gives a very faithful, albeit semiquantitative, reproduction of the results of peak integration. Pb(II) appears to have no effect on the matrix at 28 d, but we have shown in older samples that the 28-d result is a coincidence. Pb(II) promotes polymerization of the silicate anions and, at longer times of cure, the Pb-OPC samples contain a much higher proportion of Q^2 units. Both forms of As are set retarders, in that the percent hydration at 28 d is smaller than for OPC alone. They also both inhibit polymerization of silicate units, and at 28 d the proportion of Q^2 units is lower than for OPC alone. Both Cr species are also set retarders, but in a different way from the As species. Cr inhibits hydration of orthosilicate (Q^0) units, just as lead does in the early stages of hydration. However, the Cr effects, particularly that of Cr(III), are much greater than the lead effects at 28 d. The Cr(III)-OPC sample has only undergone 35% hydration at 28 d compared to almost double that for the As and Pb samples. The consequences of metal additions on the chemical structure of the matrix are obvious from the NMR results, but in most cases relationships between structure and properties remain to be explored.

A number of samples have been examined in much more detail, including NMR as a function of time, and the following account describes results with Cd- and Pb-containing cements.[12] The very basic nature of hydrating cement pastes could be expected to result in precipitation of metal hydroxides. Even though $Cd(OH)_2$ and $Pb(OH)_2$ have comparable, and very low, solubilities, the leachability of solidified products containing Cd^{2+} and Pb^{2+} is quite different. Concentrations of Cd in leachates are very low, while Pb concentrations are considerably higher.

The added Cd salts have minor effects on cement hydration. The formation of $Ca(OH)_2$ and calcium silicate hydrate "gel" (C-S-H), which are the major phases characteristic of hydrated cement pastes, is accelerated, and the "mature" paste at 28 d shows a slightly higher proportion of C-S-H and a slightly higher degree of silicate polymerization when compared to cement without Cd.

Lead salts show very different effects. It is well known that Pb salts are set retarders. That effect is very clear in the present studies, and NMR shows that retardation applies to both aluminate and silicate phases. After an extended dormant period lasting up to 3 d, normal hydration reactions begin to occur, and a 28-d sample has an appearance, both macroscopically and by SEM, that is hardly distinguishable from cement without Pb. NMR shows that both C-S-H and $Ca(OH)_2$ are forming, but both NMR and infrared studies[13] show a higher degree of silicate polymerization in the matrix. Furthermore, in contrast to cement alone, the increase in Q^2 units at the expense of Q^1 continues to the point that after a year the degree of polymerization in the matrix is quite different when Pb is present.[13]

As a model system for the effects of $Pb(NO_3)_2$ on early hydration of Portland cement, Thomas et al.[14] carried out potentiometric titrations of $Pb(NO_3)_2$ against $Ca(OH)_2 + CaSO_4$. They observed an ill-defined end point with pH fluctuating between 8.5 and 10, and explained the results in terms of formation of mixed salts of Pb containing all 3 anions and the occurrence of solid formation followed by dissolution and then reprecipitation. The latter suggestion is supported by the fact that Pb, in contrast to more well-behaved metals, appears in high concentration in surface coatings when the solids are fractured[12] or microtomed.[15]

Out of all these results, a tentative picture emerges of the differences in Cd and Pb solidification. The Cd/cement system involves early formation of $Cd(OH)_2$, which then provides nucleation sites for precipitation of C-S-H and $Ca(OH)_2$, resulting in Cd being in the form of the insoluble hydroxide with a very impervious coating. On the other hand, the Pb/cement system contains mixed Pb salts which retard cement hydration by forming an impervious coating around cement clinker grains. However, as pH in the cement pore waters undergoes fluctuations during the progress of hydration, the Pb salts undergo solubilization and reprecipitation. The result is the presence of Pb salts on the surfaces of cement minerals even in "mature" cement pastes. The Pb salts are readily accessible to leach water and are apparently more soluble under the basic conditions than a pure lead hydroxide would be.

6.4. MIXED SALT SYSTEMS

In real wastes which might be subjected to S/S, it would be unusual to have a single heavy metal component. Consequently, we have looked at a number of other simulated wastes containing more complex mixtures in which the chemistry is potentially much more complex. The environmental laboratory of the U.S. Army Corps of Engineers' Waterways Experiment Station prepared a synthetic sludge by mixing an aqueous solution of the nitrates of Cd, Cr, Hg, and Ni with lime, followed by dewatering. The resulting sludge was approximately 10% by weight in Cr and Ni, and a few tenths of a percent in Cd and Hg. The sludge was subsequently solidified in a number of binders, including OPC. We have investigated a large number of these solidified products, mainly by scanning electron microscopy (SEM), energy dispersive X-ray microanalysis (EDX), and powder X-ray diffractometry (XRD).[16]

The sludge itself underwent considerable carbonation from atmospheric CO_2 during its preparation (and perhaps afterward). XRD characterization of the sludge alone showed a number of peaks identifiable with $Ca(OH)_2$ and $CaCO_3$. There were other strong peaks in the diffraction patterns, but no simple hydroxides or carbonates of the heavy metals could be identified.

When the sludge was solidified with OPC, there was little effect on cement hydration. The powder diffraction patterns of the mixtures were the sum of the peaks associated with the sludge and with hydrated Portland cement. SEM of the samples frequently showed abundant ellipsoidal particles tens of microns across, usually separated from the surrounding matrix by a thin, void rim. Such particles are not seen in OPC samples. EDX analysis of the particles showed high concentrations of Cr and Ni, in addition to Ca, Si, Al, and S. The matrix several tenths of a millimeter from these large particles had the same elements, but with Cr and Ni in much lower concentration.

Thus, this S/S system gives every appearance of being only a physical mixture of waste and binder with very little chemical interaction between the two; essentially physical encapsulation on a microscopic scale. It is not yet clear whether the fact that the surfaces of the sludge particles were strongly carbonated contributed to this lack of chemical interaction. The fact that the waste contains a high percentage of solids may also be relevant in determining waste-matrix interactions. However, it is not necessarily a controlling factor, since in the lead work cited above, solutions of $Pb(NO_3)_2$ and "$Pb(OH)_2$" sludges had similar effects of retardation and alteration of silicate polymerization patterns.

6.5. SOLIDIFIED SAMPLES AFTER LEACHING

The TCLP (or EP Tox) leaching procedures continue to be regulatory instruments used by the U.S. EPA for classifying wastes as hazardous and for assessing the acceptability of various wastes, or solidified wastes, for land disposal. It has been appreciated for a long time that the pH of the leaching solution, most particularly the pH of the solution after leaching has been carried out, is a major variable in determining the amount of a metal that will be leached. Indeed, workers at Environment Canada have devised a leaching

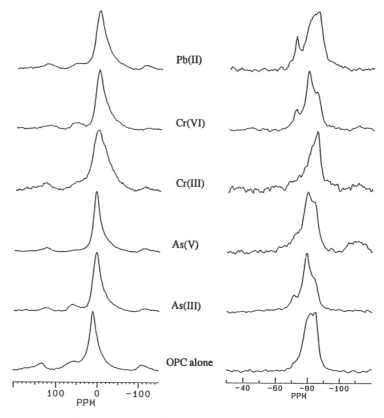

Figure 6-2. ^{27}Al **(left) and** ^{29}Si **(right) solid-state MAS NMR spectra obtained at 4.7 T of solid residues after TCLP leaching of type I Portland cement samples cured for 28 d and containing 10% by weight of As (III), As (V), Cr (III), Cr (VI), and Pb (II).**

procedure in which the pH is controlled to be 5.0 throughout the leaching process, and they have suggested that the results are indicative of the total amount of metal that is leachable under a worst-case environmental situation.[17]

The same OPC-metal salt mixtures described in Figure 6-1 have also been treated in the following way. After 28 d of cure, the samples were ground as specified in the TCLP and then leached using two protocols: the TCLP and a NaOAc/HOAc mixture with sufficient buffering capacity to maintain pH 5 throughout the extraction. After leaching, the solid residues were dried and examined by FTIR and NMR. The Al and Si NMR results for the TCLP leach are shown in Figure 6-2 and the pH 5 leach in Figure 6-3. These spectra may also be compared directly with those in Figure 6-1, which show the same samples before leaching.

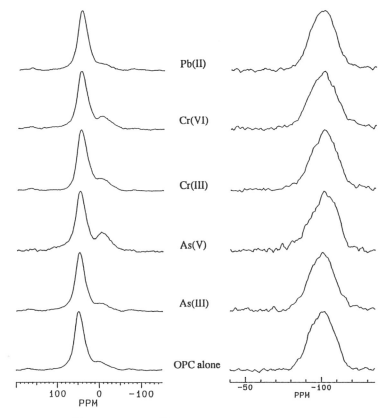

Figure 6-3. ^{27}Al (left) and ^{29}Si (right) solid-state MAS NMR spectra obtained at 4.7 T of solid residues after pH 5 leaching of type I Portland cement samples cured for 28 d and containing 10% by weight of As (III), As (V), Cr (III), Cr (VI), and Pb (II).

After leaching at pH 5, all the samples present almost exactly the same picture spectroscopically. The aluminates have been converted almost entirely back to the tetrahedrally coordinated form, although there are small variations in the tetrahedral-to-octahedral ratios. The silicates have been converted from a mixture of orthosilicate units, SiO_3^{3-} terminal units, and SiO_2^{2-} chain-lengthening units into a mixture that is similar to amorphous silica and contains only chain-branching and cross-linking silicate units. Clearly, the calcium silicate hydrate matrix that is characteristic of OPC has been completely destroyed. This transformation is accompanied by greatly enhanced metal concentrations in the leachate compared to leaching by distilled water or under TCLP conditions.

Leaching under TCLP conditions has much less dramatic effects upon the matrix. The conversion of octahedral aluminates to tetrahedral has hardl··

begun in most cases. Orthosilicate units are diminished in number and additional terminal and chain-lengthening units appear, but the extent to which orthosilicate has disappeared and the extent to which chain-branching and crosslinking units have appeared is quite dependent on the metal which has been solidified. This is true despite the fact that all the solutions after leaching have nearly the same pH; namely 11.5 to 12. Since the degree of matrix alteration that occurs under TCLP leaching is quite variable depending upon the waste involved, the logical conclusion is that TCLP leaching data cannot be interpreted in a logical manner in terms of mechanism of containment. For instance, it is not reasonable to compare two TCLP leachate concentrations from two different samples containing different metals and say that one of the metals is bound more strongly in the matrix than the other. Thus, the TCLP, while perhaps having regulatory value, does not have intrinsic value for understanding leaching behavior and its causes. On the other hand, the pH 5 leach, while perhaps poorly representative of environmental conditions that may be encountered, is nevertheless interpretable from one experiment to another in a consistent manner. It always represents the results to be expected when the cement matrix is completely destroyed.

6.6   PHENOLIC WASTES

Waste streams and contaminated soils containing both metals and organics are commonly encountered, and the applicability of S/S to any organic-containing wastes has been hotly debated. A common case of such cocontamination is that in which the organics have appreciable water solubility. As an example of that system, we have carried out a large number of studies of phenols, solidified with cementitious binders both alone and with metals. Phenol itself is a commonly encountered waste product, and hence we have looked at it in some detail along with *p*-bromophenol (pBP) and *p*-chlorophenol (pCP). The latter have been included in our studies because substituted phenols are common articles of commerce, but also because the Br and Cl atoms allow easy detection in energy dispersive X-ray (EDX) analysis. We have also shown in separate studies that Cl does not become detached from the aromatic ring during solidification in OPC;[18] hence, detection of Cl means detection of the substituted phenol.

Effects of Phenols on the Matrix

Hydroxy-substituted organics, particularly polyhydroxy compounds, are well-known cement set retarders, although the effects of the phenols are fairly small.[19] Furthermore, compressive strengths of OPC containing phenols are nearly the same as OPC alone until phenol loadings exceed 10% by weight, and even then strength losses are rather minor up to 20% by weight. SEM and XRD do show matrix changes with increasing proportions of phenols. There is decreasing crystallinity, as judged qualitatively from SEM and from the increasing background noise relative to sharp peaks in the XRD patterns. Nevertheless, even at 20% by weight phenol, both transmission electron microscopy[20] and XRD[18,21] show $Ca(OH)_2$ to still be present in the matrix. XRD shows increasing occurrence of new sharp peaks not present in cement alone. Comparison of XRD patterns of phenol-containing cements with those of Ca salts of phenols prepared separately shows peak coincidences, but enough of the salt peaks are obscured by noise or overlapping strong cement peaks that it is not possible to positively identify the salts in the phenol-cement mixture.[18]

In samples containing 10% by weight phenol, NMR shows an initial acceleration of the hydration of both aluminate and silicate phases caused by the presence of pBP and pCP.[22] However, there is a longer dormant period in the presence of the phenols than in their absence, and the matrix at 28 d contains more Q^0. After 1 year of cure, the phenol-containing matrices show silicate phases that are indistinguishable by NMR from cement alone.

Effect of Cement on the Phenol

It has been assumed in the cement literature that phenol is converted to calcium phenoxide in contact with cement clinker. Indeed a qualitative test for the presence of free lime in clinker involves treatment with phenol in nitrobenzene and water and microscopic observation of the characteristic long needles of "calcium phenoxide".[23] Phenol and monosubstituted phenols typically have pKa's in the range 9 to 11, and the pH of pore waters in cements is very basic, up to about pH 12.5. Consequently, the potential for conversion of phenols to their anions exists, but the extent of conversion may depend upon a number of factors controlling the acidity of the phenol and the basicity of the pore waters. The cation present in overwhelming amount is Ca^{2+}, consequently a Ca phenoxide would be expected. It has recently been shown[24] that combination of equimolar quantities of $Ca(OH)_2$ and PhOH produces PhOCaOH, not $(PhO)_2Ca$, and the latter can only be formed under forcing conditions with continuous removal of water. The OH group of PhOCaOH shows a characteristic IR signal due to non-hydrogen bonded OH, as does $Ca(OH)_2$ (but with different wave number). Also, it is possible to distinguish ionized from nonionized phenol by solid-state ^{13}C NMR, since the *ipso*-carbon of the aromatic ring shifts upfield by about 7 ppm in either PhOCaOH or $(PhO)_2Ca$ compared to PhOH.[24]

We have used both FTIR and ^{13}C NMR to investigate samples of solidified cement containing 10% PhOH, pBP, and pCP.[25] pBP and pCP also form the phenoxycalcium hydroxide on treatment of the phenol with $Ca(OH)_2$, as shown by a characteristic OH peak. However, it is not possible to tell by IR which form the phenols are in after solidification with cement, since the strong

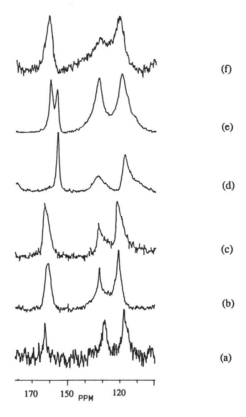

Figure 6-4. ^{13}C solid-state NMR spectra (CPMAS at 4.7 T) of phenols in solidified type I Portland cement (OPC) containing: (a) 10% phenol, cured 2 months; (b) 10% pBP, cured 1 year; (c) 10% pCP, cured 1 year; (d) 10% pBP + 10% Cd from Cd(NO$_3$)$_2$, cured 7 d; (e) 10% pBP + 10% Pb from Pb(NO$_3$)$_2$ cured 3 d; (f) 10% pBP + 10% Cd from Cd(NO$_3$)$_2$, cured 2 months.

absorptions due to the cement matrix are overwhelming. The fact that there are XRD peak coincidences[18] suggests that ArOCaOH is present, but is not conclusive. ^{13}C NMR shows that there is considerable variation in the proportions of ionized and nonionized phenol in the matrix depending on the length of time of cure and the presence of heavy metals along with phenol. As shown in Figure 6-4, OPC with 10% PhOH, pBP or pCP contains only ionized phenols. However, with Cd or Pb salts present, some (case e) or all (case d) of the phenol may be present in nonionized form, and the proportion also appears to be a function of the time of cure.

Some time ago we reported data on extractions of ethylene glycol from pulverized solids prepared from Portland cement and 10% by weight glycol.[26]

That work used pure solvents of varying polarities and also multiple extractions. We concluded that the glycol appeared to occupy more than one environment because polar, aprotic solvents like dimethyl sulfoxide extract much less glycol than water and because multiple extractions even with water did not account for 100% of the glycol. Similar results are obtained for pBP.[22] Water extracts a relatively high proportion of phenols (40 to 60%) from cement-solidified samples, but the proportion decreases substantially between 90 d and 1 year of cure — down to 9% recovery for pBP at an original loading of 10% by weight.

We have recently used a much more sophisticated solid-state NMR technique to explore the extent to which phenols may occupy tightly bound environments in cement matrices.[27] Deuterium NMR spectra and spin-lattice relaxation times (T_1) can be simulated for many molecular reorientational modes and rates. In our work, spectra have been acquired for d_5-phenol itself and for samples containing 0.1, 1, and 10 wt% of the phenol in cement. The dominant feature for all these samples between 260 and 360 K is a liquid-like spectrum, although at lower temperatures a static line shape is observed. Tightly bound phenol cannot be seen in these spectra because of the dominance of the contribution from loosely bound phenol. However, the latter can be removed by crushing and either drying the sample in an oven (36 h at 90°C) or allowing evaporation at room temperature over 30 d. For either kind of sample, the deuterium spectrum now shows reduced spectral intensity, a line shape corresponding to 180° flips of the aromatic rings, and little evidence of a liquid-like phase. The quality of the T_1 data is not sufficient to allow us to determine the activation energy for ring flips unambiguously, and some samples give evidence that there is more than one component, i.e., phenol in more than one kind of tightly bound environment. Our best estimate of these activation energies lies in the range of 6 to 9 kcal/mol. This is an important number, as it sets a lower limit for the phenol-cement bond strength, since the activation energy for ring flipping must be less than the dissociation energy. Our current physical interpretation of these results is that most of the phenol is dissolved (in most cases as the phenoxide ion, although in some cases as nonionized phenol as well) in the pore water of the cement matrix. We assume that our crushed samples allow evaporation of water and eventual crystallization of the phenoxide as (mainly) Ca salts.

We are slowly developing a rather sophisticated picture of phenol-cement interactions, but there are clearly subtleties that may defy explanation for some time to come. One of those presumably arises because phenols may be ionic or nonionic or both, and a number of factors contribute to the distribution among these forms. Indeed, in comparisons of cement samples containing pBP and pCP, we can see obvious, gross differences. Examination under the optical microscope shows the presence of phases in pBP-cement samples that are tens of microns across and that are not present in pCP-containing samples. Electron probe microanalysis shows high concentrations of Br in these phases and low concentrations in the matrix. On the other hand, corresponding analysis of pCP-containing samples shows homogeneous distribution of Cl on that physical scale. It is tempting to conclude that the Ca phenoxide crystallizes as a distinct phase in the pBP samples, but is widely distributed throughout the C-S-H phase in the pCP samples. It may be that factors as subtle as the atomic size difference between Cl and Br result in very different behavior despite the fact that the pKa's of the two substituted phenols are not much different.

6.7. CONCLUSIONS

This paper has emphasized the methodology available to gain chemical information about solidified wastes and a survey of results in our group to date. We have proceeded from the premise that it is necessary to gain basic chemical data before attempting to understand or predict performance. Furthermore, we have used simple systems in the beginning in order to be confident that we understand and correctly interpret the instrumental output we are accumulating. This work has already provided us with some rather practical insights into the process of S/S and its evaluation that can be applied now. We have information about what the TCLP is doing and what limitations there are to its interpretation. We have clearly shown that certain admixtures result in matrix changes occurring over long periods of time — an observation that clearly points out the need for long-term testing during evaluation of S/S. It is obvious from our work that the conventional wisdom is not correct when it says that the effects of certain set-retarding chemicals are short-term effects only.

However, we have also generated a great deal of information that is not so easy to interpret in terms of S/S practice. We have shown clearly that Cr^{3+} greatly retards the normal polymerization process of orthosilicate units, so that "mature" (28 d of cure) cement has undergone a minor degree of hydration. That effect, however, is not relevant to the effectiveness of Cr(III) immobilization, since the Cr is not readily leached. Probably that means that the Cr is immobile mainly because it has been converted into a rather insoluble salt (or salts, presumably a hyroxide or mixed salt containing hydroxide), and the matrix is making little contribution to the Cr immobilization. However, what are the long-term effects of the matrix differences on Cr immobilization or on physical properties of the solidified material? We do not know. In a mixture of Cr(III) with other metals, does Cr still have the same effects, and do the Cr-dependent matrix changes affect the immobilization of the other metals? We do not know that either.

We now know the effects which various metals and organics have on the degree of polymerization of the silicate units of the cement matrix used in S/S. Is the more highly polymerized matrix a better medium for immobilization, or a worse one? There is direct evidence that increasing degree of hydration, and particularly increasing proportion of Q^2 units, results in increased compressive strength in cement pastes,[28] but compressive strength is only one indicator of performance.

We can determine whether phenols are ionized or not after solidification, and it would probably be possible to shift the equilibria toward one side or the other with appropriate additives to the S/S mix. What form of the phenol do we want in order to optimize the process? From the point of view of phenol immobilization, most chemists would be likely to respond that the ionized form would be more likely to be mobile in aqueous media and therefore less desirable. However, we find that long times of cure result in both more ionized phenols and in lower leachabilities. The Ca salts (and there may be several of them involved) may be less soluble in water than the nonionized phenol.

Clearly, we have a long way to go to gain very complete understanding of the complex systems involved in S/S. It is also always appropriate to ask, "Is that piece of knowledge relevant to S/S practice?" Nevertheless, we have shown that a great deal of chemical information is accessible, and we have shown that some of it is useful. We are confident the potential exists for conversion of S/S practice from an art to a science.

ACKNOWLEDGMENTS

This work has been generously supported by the U.S. EPA through direct contracts with the Risk Reduction Engineering Laboratory, Cincinnati; Charles R. Mashni and Carlton C. Wiles, project managers; and through U.S. EPA-supported centers at Louisiana State University (Hazardous Waste Research Center) and Lamar University (Gulf Coast Hazardous Substance Research Center). The authors are all faculty at Louisiana State University, but gratefully acknowledge that the bulk of the work reported herein has been done by our students, who are too numerous to include in the list of authors, but who may be identified in the publications contained in the list of references.

REFERENCES

1. Cartledge, F. K. Author to whom correspondence should be addressed at the Department of Chemistry. Louisiana State University, Baton Rouge, LA 70803-1804.

2. Conner, J. C. 1990. *Chemical Fixation and Solidification of Hazardous Wastes*. Van Nostrand Reinhold, New York.

3. Lippmaa, E., M. Mägi, A. Samoson, G. Engelhardt, and A.-R. Grimmer. 1980. "Structural Studies of Silicates by Solid-State High-Resolution ^{29}Si NMR." *J. Am. Chem. Soc.* 102: 4889–4893.

4. Fyfe, C. A., J. M. Thomas, J. Klinowski, and G. C. Gobbi. 1983. "Magic Angle Spinning NMR (MAS-NMR) Spectroscopy and the Structure of Zeolites." *Angew. Chem. Intern. Ed. Engl.* 22: 259–275.

5. Lippmaa, E., M. Mägi, M. Tarmak, W. Wieker, and A.-R. Grimmer. 1982. "A High Resolution ^{29}Si NMR Study of the Hydration of Tricalcium Silicate." *Cem. Concr. Res.* 12: 597–602.

6. Clayden, N. J., C. M. Dobson, G. W. Groves, C. J. Hayes, and S. A. Rodger. 1984. "Solid State NMR Studies of Cement Hydration." *Br. Ceram. Proc.* 35: 55–64.

7. Barnes, J. R., A. D. H. Clague, N. J. Clayden, C. M. Dobson, and C. J. Hayes. 1985. "Hydration of Portland Cement Followed by ^{29}Si Solid-State NMR Spectroscopy." *J. Mater. Sci. Lett.* 4: 1293–1295.

8. Muller, D., A. Rettel, W. Gessner, and G. Scheler. 1984. "An Application of Solid-State Magic-Angle Spinning ^{27}Al NMR to the Study of Cement Hydration." *J. Mag. Reson.* 57: 152–156.

9. Lahajnar, G., R. Blinc, V. Rutar, V. Smolej, I. Zupancic, I. Kocuvan, and J. Ursic. 1977. "On the Use of Pulse NMR Techniques for the Study of Cement Hydration." *Cem. Concr. Res.* 7: 385–394.

10. Schreiner, L. J., J. C. MacTavish, L. Miljkovic, M. M. Pintar, R. Blinc, G. Lahajnar, D. Lasic, and L. Reeves. 1985. "NMR Line Shape — Spin Lattice Relaxation Correlation Study of Portland Cement Hydration." *J. Am. Ceram. Soc.* 68: 10–16.

11. MacTavish, J. C., L. Miljkovic, M. M. Pintar, R. Blinc, and G. Lahajnar. 1985. "Hydration of White Cement by Spin Grouping NMR." *Cem. Concr. Res.* 15: 367–377.

12. Cartledge, F. K., L. G. Butler, D. Chalasani, H. C. Eaton, F. P. Frey, E. Herrera, M. E. Tittlebaum, and S.-L. Yang. 1990. "Immobilization Mechanisms in Solidification/Stabilization of Cd and Pb Salts Using Portland Cement Fixing Agents." *Environ. Sci. Technol.* 24: 867–873.

13. Ortego, J. D., Y. Barroeta, F. K. Cartledge, and H. Akhter. 1991. "Leaching Effects on Silicate Polymerization — An FTIR and ^{29}Silicon NMR Study of Lead and Zinc in Portland Cement." *Environ. Sci. Technol.* 24: 1171–1174.

14. Thomas, N. L., D. A. Jameson, and D. D. Double. 1981. "The Effect of Lead Nitrate on the Early Hydration of Portland Cement." *Cem. Concr. Res.* 11: 143–153.

15. Cocke, D., J. D. Ortego, H. McWhinney, K. Lee, and S. Shukla. 1989. "A Model for Lead Retardation of Cement Setting." *Cem. Concr. Res.* 19: 156–159.

16. Roy, A., H. C. Eaton, F. K. Cartledge, and M. E. Tittlebaum. 1991. "Solidification/Stabilization of a Heavy Metal Sludge by a Portland Cement/Fly Ash Binding Mixture." *Haz. Waste Haz. Mater.* 8: 33–41; Roy, A., H. C. Eaton, F. K. Cartledge, and M. E. Tittlebaum. 1992. "Solidification/Stabilization of Hazardous Waste: Evidence of Physical Encapsulation." *Environ. Sci. Technol.* 26: 1349–1353; Roy,. A., H. C. Eaton, F. K. Cartledge, and M. E. Tittlebaum. 1992. "The Effect of Sodium Sulfate on Solidification/Stabilization of a Synthetic Electroplating Sludge in Cementitious Binders." *J. Haz. Mater.* 30: 297–316.

17. Coté, P. L. 1989. "Proposed Evaluation Protocol for Stabilized/Solidi-fied Wastes." Wastewater Technology Center, Environment Canada; U.S. EPA Office of Research and Development; and Ontario Ministry of the Environment, Waste Management Branch.

18. Cartledge, F. K., H. C. Eaton, and M. E. Tittlebaum. 1990. "The Morphology and Microchemistry of Solidified/Stabilized Hazardous Waste Systems." U.S. EPA Project Report (EPA/600/S2-89/056).

19. Sheffield, A., S. Makena, M. Tittlebaum, H. Eaton, and F. Cartledge. 1987. "The Effects of Three Organics on Selected Physical Properties of Type I Portland Cement." *Haz. Waste Haz. Mater.* 4: 273–286.

20. Chou, A. C., H. C. Eaton, F. K. Cartledge, and M. E. Tittlebaum. 1988. "A Transmission Electron Microscopic Study of Solidified/Stabilized Organics." *Haz. Waste Haz. Mater.* 5: 145–153.

21. Walsh, M. B., H. C. Eaton, M. E. Tittlebaum, F. K. Cartledge, and D. Chalasani. 1986. "The Effect of Two Organic Compounds on a Port-land Cement-Based Stabilization Matrix." *Haz. Waste Haz. Mater.* 3: 111–123.

22. Chalasani, D. 1988. "Development of Techniques for Studying Interac-tions of Organic Compounds with Complex Solid Matrices." Ph.D. dissertation, Louisiana State University, Baton Rouge.

23. Lea, F. M. 1971. *The Chemistry of Cement and Concrete,* 3rd Ed. Chemical Publishing, Chicago.

24. Schlosberg, R. H. and C. G. Scouten. 1988. "Organic Chemistry of Calcium. Formation and Pyrolysis of Hydroxycalcium Phenoxides." *Energy Fuels* 2:582–585.

25. Akhter, H. and F. K. Cartledge. 1990. Unpublished studies.

26. Chalasani, D., F. K. Cartledge, H. C. Eaton, M. E. Tittlebaum, and M. B. Walsh. 1986. "The Effects of Ethylene Glycol on a Cement-Based Solidification Process." *Haz. Waste Haz. Mater.* 3: 167–173.

27. Janusa, M. A., X. Wu, F. K. Cartledge, and L. G. Butler. 1992. "Solid-State Deuterium NMR Spectroscopy of d_5-Phenol in White Portland Cement." manuscript in preparation.

28. Parry-Jones, G., A. J. Al-Tayyib, S. U. Al-Dulaijan, and A. I. Al-Mana. 1989. "^{29}Si MAS-NMR Hydration and Compressive Strength Study in Cement Paste." *Cem. Concr. Res.* 19: 228–234.

7 MICROSTRUCTURAL CHARACTERIZATION OF CEMENT-SOLIDIFIED HEAVY METAL WASTES

D. L. Gress and T. El-Korchi

7.1. INTRODUCTION

Solidification/stabilization (S/S) is considered a technology available for treating selected waste prior to land disposal. This technology is also being considered for treating residues from other treatment technologies. S/S processes employ select materials to alter the physical and chemical characteristics of the waste stream prior to land disposal. Reviews of S/S technology for hazardous industrial wastes have been recently published.[1-12]

Most inorganic fixation processes use some type of hydraulic cement, pozzolanas, lime, gypsum, and soluble silicates. Portland cement is most common; however, other types have also been used such as aluminous cement, natural cement, slag cement, etc. Cement-based S/S processes are well suited for treatment of aqueous waste since Portland cement and pozzolonic materials require water for their hydration. The leachability of the containment thus depends on whether it remains in solution in the pore system or is immobilized through chemical reaction. Understanding the containment mechanism responsible for the S/S process is very useful in providing information on potential release of metals.

7.2. OBJECTIVE

A better understanding of the fixation mechanism is necessary for improved design processes that would ultimately reduce pollutant mobilization in S/S

0-87371-748-1/93/$0.00+.50

Table 7-1. Concentration of Metal in Sludge

Metal	Concentration	
	ppm	M
Arsenic	24,000	0.32
Cadmium	23,000	0.20
Chromium	24,000	0.46
Lead	23,000	0.11
Mixed metals		
Arsenic	3,000	0.04
Cadmium	4,500	0.04
Chromium	2,080	0.04
Lead	8,290	0.04

wastes placed in landfill environments. Type of fixation is highly dependent on microstructure development of the cement S/S-based wastes. To achieve this goal, the laboratory investigation utilized the following methods to help characterize the microstructure development: X-ray diffraction analysis, scanning electron microscopy (SEM), energy-dispersive spectrometry, mercury intrusion, and helium displacement porosimetry.

7.3. EXPERIMENTAL

Sample Preparation

The contaminants cadmium, lead, chromium, and arsenic were selected for their common occurrence in industrial wastes and their high toxicity characteristics, The solidified wastes were prepared with synthesized heavy metal sludges and a type II Portland cement in accordance with Reference 13.

The heavy metal sludges were prepared from cadmium nitrate, chromium nitrate, lead nitrate, and sodium arsenite. The solution pH was adjusted to 8.5 with 6.0 M sodium hydroxide. Partial precipitation of the cationic metals occurred at this pH; however, arsenic was added as the anion, sodium arsenite, and did not precipitate at pH of 8.5 The metal concentrations after pH adjustment are shown in Table 7-1.

These heavy metal sludges were mixed with type II Portland cement. The sludges were compensated for solids content with distilled water to obtain a water-cement ratio of 1.0. Sample cylinders measuring 1.5 by 3.0 in. were molded and left to set overnight before stripping and curing in a humidity room with a relative humidity of approximately 100%.

Table 7-2. QXRD Analysis Relative Intensities Ratios

Cadmium to cement	1.0^a	3.0	5.0
Calcium hydroxide to cadmium	5.58^b	3.67	1.04

[a] By weight.
[b] Relative intensities.

Procedure and Results

X-Ray Diffraction Analysis

Precipitates in the metal sludges were analyzed to determine their crystalline compound forms in an effort to possibly locate them in the cement waste matrix. The metal hydroxide precipitate should be a stable form in the alkaline environment of the paste. The lead sludge was identified as lead oxide nitrate hydroxide [$Pb_6O_3(NO_3)_2(OH)_4$]. The cadmium sludge precipitated as cadmium hydroxide [$Cd(OH)_2$]. Chromium did not exhibit any strong diffraction peaks; however, there were broad peaks between 37.5 and 39.8° 2 theta. The arsenite solution did not precipitate at pH of 8.5 and X-ray diffraction was not possible; however, the solution was oven dried and as expected sodium arsenite crystallized [$NaAsO_2$].

XRD analysis was conducted on all wastes after 7 and 28 d and 3 years of hydration. Cadmium hydroxide (CH) was the only crystalline metal hydroxide detected in all the metal wastes. The effect of different metal sludges on the peak intensity of major crystalline compounds in partially hydrated Portland cement was observed by El Korchi.[13] Qualitative changes in the hydrous and anhydrous phases were observed. However, due to their crystallinity, only the cadmium waste was investigate quantitatively.

The effect of cadmium on the hydration of Portland cement is investigated by quantitatively monitoring the calcium hydroxide formation as a function of the cadmium content added to the mix. The ratio of peak intensities is used since the ratio of peak intensities of two substances is directly proportional to their weight percentage ratio. The intensity ratios are obtained by measuring the area under the respective peaks; details are formed in Reference 13. The data are presented in Table 7-2.

Scanning Electron Microscopy and Energy-Dispersive Spectrometry

SEM was utilized to observe the microstructure of the different S/S wastes during the temporal hydration process. Energy-dispersive spectrometry was

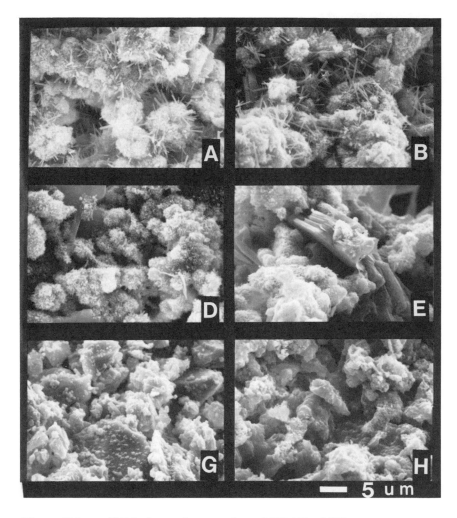

Figure 7-1.   SEM photomicrograph and EDAX of S/S wastes:

Sample	3 d	7 d	3 Years
Control	A	B	C
Cadmium	D	E	F
Lead	G	H	I

used to locate the heavy metals within the hydrated cement matrix and identify the various microstructures by determining their elemental composition.

The microstructure of the waste observed under the SEM is shown in Figures 7-1 and 7-2. These SEM photomicrographs show typical microstructures at 3, 7, and 28 d and 3 years of hydration accompanied by a typical EDAX spectrum at 3 years.

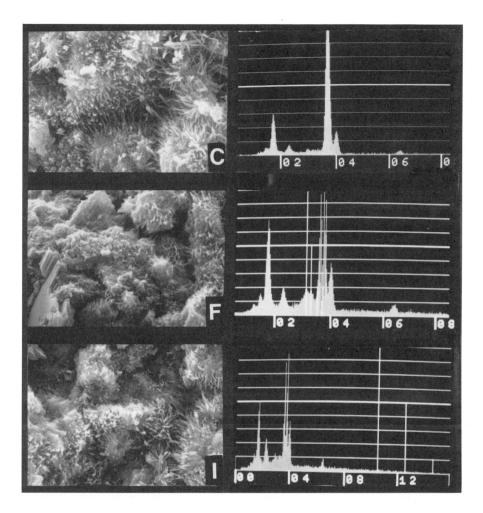

Figure 7-1 continued.

Microstructure of the control paste specimen is shown in Figure 7-1 A to C. The microstructure of 3-d hydration is typified by anhydrous cement grains coated with a hydrated calcium silicated hydrate (CSH) gel layer with initial fibrous morphology. Ettringite crystals appear prismatic and deposit in the voids or on the surface of hydration material. Calcium hydroxide crystals initiate as thin hexagonal plates and then develop into larger striated ones induced by fracture during sample preparation. Hydration observed in the specimens cured for 3 years resulted in continued CSH development and interconnecting cement particles. CH and Aft crystal grow larger as well.

Microstructural differences are observed upon addition of the heavy metals. The cadmium waste (Figure 7-1 D to F) exhibits a more reticulated and fibrous

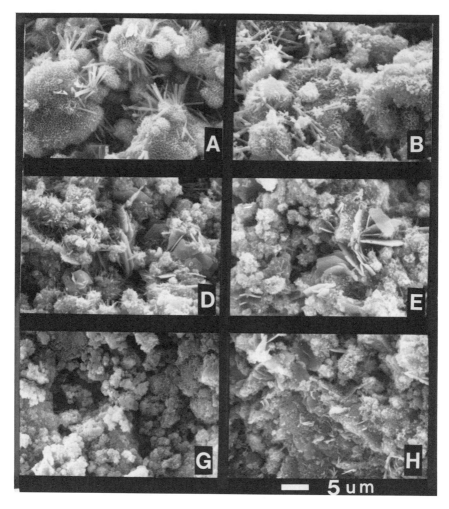

Figure 7-2.   SEM photomicrograph and EDAX of S/S wastes:

Sample	3 d	7 d	3 Years
Arsenic	A	B	C
Chromium	D	E	F
Mixed metals	G	H	I

hydration layer at 3 d and with lesser ettringite crystals as compared to the control. The growth of CH crystals appears normal. Continued hydration for 3 years exhibits typical microstructures. The lead waste (Figure 7-1 G to I) exhibits the most dramatic difference in early hydration characteristics. The 3-d specimens showed only anhydrous cement grains with thin gelatinous

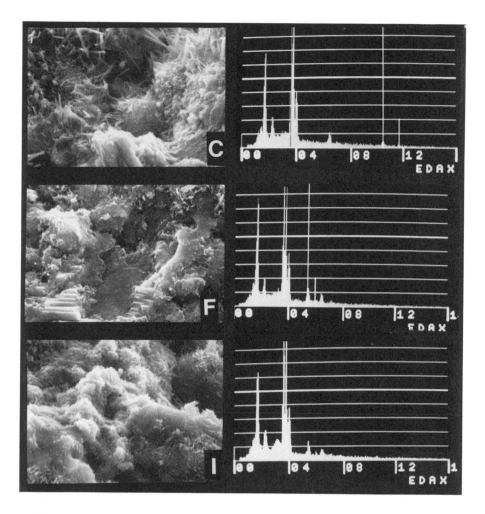

Figure 7-2 continued.

surfaces at 7 d. After continued hydration into 28 d and longer, typical hydration microstructure appears.

Figure 7-2 consists of representative SEM photomicrographs for arsenic (Figure 7-2 A to C), chromium (Figure 7-2 D to F), and mixed metals (Figure 7-2 G to I). The addition of sodium arsenite creates a highly reticulated surface at early ages and well-developed ettringite crystals. Continued hydration interconnects the cement grains with the fibrous hydrates and well-developed Aft. Calcium hydroxide is not apparent at 3 d but is well developed at 28 d and 3 years.

The chromium waste exhibits well-developed CH hexagonal platelets; however, it was not possible to determine if the smaller thin plates are Afm crystals.

The hydration produced, covering the cement grains, appears to be equant grain and not as fibrous as observed previously. Continued hydration shows well-developed CH plates and visibly denser material.

The mixed metal waste has a microstructure similar to anhydrous cement grains with a thin layer of hydration product. After 7 d of hydration, hexagonal platelets appear with minute fibers. The mature paste at 3 years shows a visibly dense structure with equant grain CSH and CH plates.

Chemical analysis using EDAX was performed on various microstructures as observed by SEM to qualitatively determine their association with different metals. X-ray dot maps were made of the desired elements of interest on fractured and polished specimen wastes as shown with the cadmium waste in Figure 7-3. All waste samples appear to be homogeneously distributed within the waste matrix. However, it is difficult to associate a particular metal with a known morphology using X-ray dot mapping due to limitations associated with sample roughness and therefore beam spread.

Mercury Intrusion and Helium Displacement Porosimetry

The total porosity and pore size distribution of a solid affects its physiomechanical properties and permeability characteristics, The correlation between porosity and permeability is dependent on total porosity, pore size distribution, tortuosity of the pores, and the morphology and configuration of the pores. Total porosity and pore size distribution may be used as potential indicators of the physiomechanical properties and leaching potential of S/S wastes.

Mercury intrusion porosimetry (MIP) is very informative in determining pore size distribution, density, and specific surface area of porous materials. The pore size distribution is approximated by estimating the intruded volume of mercury under pressure.

The pore size distribution is affected by water-cement ratio and degree of hydration. Strength and permeability are governed by large pores while the small pores influence the drying shrinkage and creep.[14] Goto and Roy[1] and Mehta[14] suggest that pores larger than 50 to 100 nm may be detrimental to strength and permeability. Mehta and Manmohan[16] found a positive correlation between the permeability coefficient and volume of pores greater than 1320 Å.

Conventional tests using water for determining specific gravity/density measurements may not be suitable for S/S waste testing due to potential dissolution of waste and binder material. To measure true density of samples, full displacement of pore volume is required and alternative methods such as mercury intrusion and gas displacement may be used.

Results as presented in Figure 7-4 show the pore size distribution of a waste sample. All samples have the same characteristic distribution with the heavy metal waste specimens having the higher intruded volume compared to the

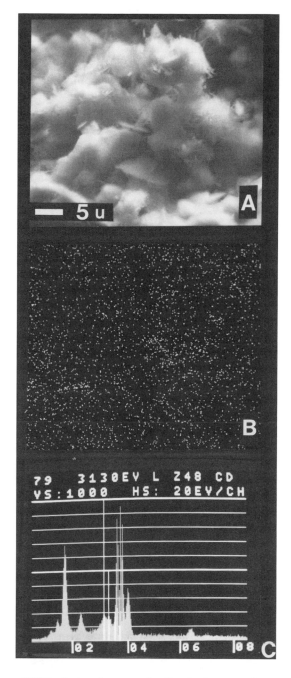

Figure 7-3.   SEM photomicrograph of cadmium waste cured 28 d: (A) secondary image, (b) X-ray dot map, and (c) EDAX spectrum.

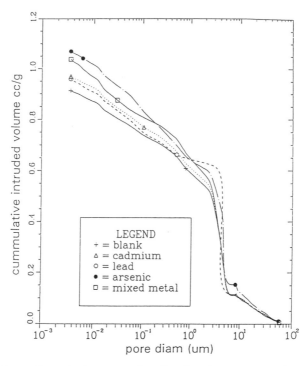

Figure 7-4.   Cumulative pore size distribution mercury intrusion.

control paste as shown in Figure 7-4. The threshold capillary (Tc), defined as the diameter where significant mercury intrusion first occurs, is approximately the same for all specimens and is located between 4 and 7 μm. The total intruded volume follows the trend arsenic > mixed metal > cadmium > lead > blank.

Hughes[17] identified the derivative of the pore size as represented by the (dV/d log r) distribution to be a sensitive indicator of pore refinement. The pore size distribution as characterized by (dV/d log h) curves has a characteristic model distribution with four peaks. The peaks are located at approximately 500, 50, 15, and 5 nm.

The differential curve is markedly affected by heavy metal addition and changes in water-cement ratio. Figure 7-5 shows the differential curve for all heavy metal and control samples.

The results for total porosity, bulk, and true density determined by the helium displacement technique[13] are presented in Table 7-3. The addition of heavy metals to the cement paste increased the porosity. The extent of this increase depends on the individual metal sludge. The trend for total porosity for specimens originally mixed at a water-cement ratio of 1.0 is as follows: chromium > arsenic > mixed metal > cadmium > lead > blank. Obviously, the

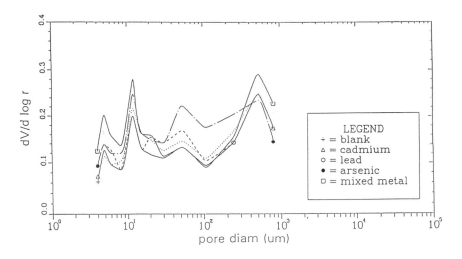

Figure 7-5.   **Derivative pore size distribution mercury intrusion.**

Table 7-3. Helium Displacement Porosimetry

Sample	Bulk Density (BD)	True Density (TD)	Moisture Content (MC)	1-MC	Porosity[a]
Blank	1.78	2.40	0.250	0.750	0.444
Lead	1.71	2.66	0.283	0.717	0.539
Cadmium	1.64	2.48	0.313	0.687	0.546
Arsenic	1.61	2.70	0.375	0.625	0.627
Chromium	1.59	2.90	0.356	0.644	0.647
Mixed metals	1.63	2.91	0.286	0.714	0.600

[a] $\text{Porosity} = 1 - \dfrac{\text{Bulk density}}{\text{True density}}[1 - \text{Moisture}]$

higher the water-cement ratio the higher the porosity as observed for the control and cadmium specimens.

Unconfined Compressive Strength

The unconfined compressive strength results and statistical analysis of the S/S waste are presented in Table 7-4. After 3 years of curing in 100% relative humidity the samples containing chromium, lead, and mixed metals all had

Table 7-4. Unconfined Compressive Strength at 14 d and 3 Years (psi)

| Sample | 14 d | | 3 Years | |
	Mean	Standard Deviation	Mean	Standard Deviation
Blank	1520	71	2197	380
Lead	1420	240	2310	266
Cadmium	870	108	1080	175
Arsenic	370	23	690	255
Chromium	1560	63	1950	255
Mixed metals	2040	100	2247	338

average compressive strengths that were not significantly different from the control specimens.

7.4. DISCUSSION

The microstructure as determined by both techniques, XRD and SEM/EDAX, had a good correlation in identifying the different microstructures during the course of temporal hydration. XRD was successful in identifying the crystalline phases in the hydrated cement waste as has been reported by Neuwirth et al.[18] Cadmium hydroxide was identified as the crystalline metal phase; however, lead, chromium, and arsenic did not exhibit similar crystalline behavior. It is not clear why the other metals did not develop a crystalline phase in the highly alkaline environment of hydrating cement. A possible explanation for this observation could be related to the nature of the precipitate forming from two supersaturated solutions.[19] These heavy metals have to form into some amorphous phase with or without other cement constituents because they do not crystallize upon drying of the sample.

The retarding effect of lead is well documented and was observed in this study. A gelatinous coating was observed around cement grains during the early periods of hydration and CH peaks were not detected in XRD analysis. This retarding or gelatinous membrane effect was not observed with chromium sludge or arsenite but slightly with cadmium. However, the loss in binding characteristic of the cadmium waste with the addition of cadmium sludge is attributed to the hindrance of calcium hydroxide formation by the cadmium hydroxide.

The Cd^{2+} ions compete with Ca^{2+} for OH^- ions. The formation of cadmium hydroxide limits the formation of calcium hydroxide. This maintains a solution supersaturated with Ca^{2+} ions and prevents further dissolution of calcium silicates which in turn limits the formation of CSH. Calcium hydroxide crystals are considered to be nucleation sites for CSH formation.[20] The limited formation of CSH ion the waste will result in a poor binding matrix.

Detection of all heavy metals was made possible by EDAX analysis. The heavy metals appear to be homogeneously distributed within the hydrated paste matrix as observed by X-ray dot maps. Spot analysis was able to detect different metals within morphologically recognized microstructure; however, it is difficult to associate a single metal with a particular hydration product due to the resolution of the SEM and contributions from other phases due to beam spread.

There is good agreement among total porosity determined by the helium displacement method, total pore volume by mercury intrusion, and the true water-cement ratio. Chromium, arsenic, and mixed metal S/S waste all have similar total porosities and mercury intruded volumes; however, the compressive strength for chromium and mixed metals is approximately three times higher than for arsenic waste. Similar behavior is observed with cadmium and lead wastes.

The relatively high characteristic compressive strength of the chromium and mixed metal specimens are due to the chlorides present in the chromium sludge used for preparing the specimens. Chromium chloride was used to make chromium sludge. The presence of chlorides would explain the high initial strength at 14 d[21] and the resulting strength at 3 years albeit at a considerably higher porosity. The higher compressive strength observed with the mixed metal samples at 14 d could be attributed to a more optimum chloride content and total porosity combination.

The arsenic sample shows considerably lower compressive strength and higher porosity. The addition of arsenic (as sodium arsenic) caused rapid set to occur within 20 min. It is not certain if this is a "flash set" resulting from insufficient soluble sulfates in pore solution. If so, it is possible that the sodium may remove sulfate ions from solution by forming sodium sulfate and the arsenic could form a calcium-arsenite complex. Conversely, the arsenic could increase the reactivity of the C_3A. The results of a "flash set" are deleterious to the potential strength gain of Portland cement paste.[14,22,23]

Since the compressive strength does not appear to correlate with porosity in all waste specimens, the chemical composition of the binding material formed must contribute to strength properties. The chemical composition of the resulting binding material in S/S wastes is just as important for mechanical properties as are the physical characteristics. The low compressive strength observed with the cadmium sample could be a result of the cadmium hydroxide formation in the hydrated phase. As previously discussed it is possible that the cadmium and calcium cations compete for the hydroxide ions to form cadmium hydroxide and calcium hydroxide. This would deplete the pore solution of OH⁻ ions, lower the pH, and weaken the CSH phase. This cadmium hydroxide formation results in reduction of calcium hydroxide formation which is proposed to be the nucleus for CSH formation.[20] Similarly, the membrane formation associated with retardation and hydration of Portland cement as discussed previously may form permanently with these cadmium samples as observed temporarily with lead.

The compressive strength of lead appears unaffected as compared to the control and may even be improved if total porosity is considered. It is possible that the increased gel formation as observed in this study and by others and the altered chemical composition of hydrates contributes to the improved strength.

The change in porosity and pore size distribution with higher water-cement ratio is expected. The higher water-cement ratio will reach a higher degree of hydration accompanied by the inherent porosity.

Convective transport through a solid matrix increases with higher porosities. A matrix with a lower porosity will provide better resistance to water and solution infiltration which is a result of the reduced permeability. This general observation does not hold for these particular S/S wastes when leached in acidic solutions[24] and seawater.[25] Shively, Melchinger, and co-workers used similar specimens for their metal leaching studies as the ones used in this study and observed different leaching behaviors depending on the metal waste. The cadmium and leads wastes have similar porosities as observed in this study; however, their release behavior was different. Cadmium release occurred earlier and at a higher rate than lead. Similar observations were made with arsenic and chromium wastes. Therefore, the release of heavy metals from the solidified matrix is more dependent on the precipitation/desorption characteristics than porosity and pore size distribution. In fact, the porosity of the matrix is continually changing during the leaching process and does not attain steady-state conditions.

7.5. CONCLUSIONS

Cadmium was found to be very prominent in combining with hydroxide to form crystalline $Cd(OH)_2$. The other solidified waste containing arsenic, chromium, and lead were not found to contain crystalline components other than those routinely associated with the formation of calcium silicate hydrate (CSH). The amorphous characteristics of the metals is related to the high pH of the supersaturated pore water inherent in the development of CSH. An amorphous precipitate of the metals within the matrix pore system and/or the CSH matrix is proposed.

The retarding and accelerating effect of a metal waste is easily evaluated using the developed techniques. A gelatinous coating observed around cement grains in conjunction with the lack of calcium hydroxide peaks during the early periods is typical of metals which retard hydration as was found with lead.

Heavy metals affect the pore system of CSH which in turn has a very pronounced affect on the physical properties of the solidified matrix (such as strength, volumetric stability, and permeability). Pore volume was investigated by mercury intrusion porosimetry and indirectly by water-cement ratio and unconfined compressive strength. Excellent agreement was found between true water-cement ratio and total pore volume by mercury intrusion. Pore size was found to be shifted toward larger pores with some of the metal wastes, thus, partially explaining the poor relationship between total poor volume and strength. Two solids of equal pore volume and different pore size distributions do not yield the same strength.

The results of this study show the addition of heavy metal wastes affects the normal temporal hydration process of Portland cement; XRD, SEM, and EDAX techniques were effective in identifying and following temporal hydration changes which in turn helped explain the mechanism of solidification; a positive correlation between total porosity and true water-cement ratio exists; and unconfirmed compressive strength is not totally dependent on total pore volume and/or pore size distribution for solidified metal wastes.

Both the SEM and XRD techniques were successful in identifying the various microstructures induced by the presence of the heavy metal wastes during the course of temporal hydration. Crystalline phases present in the hydrated matrix were effectively identified and followed during their development. A very useful technique of evaluating and monitoring the effect of a given metal waste component was developed by quantitatively monitoring the crystalline formation of CSH-related components as a function of a given metal. The ratio of peak intensities of such components obtained from XRD and EDAX is directly proportional to their weight percentages. Energy-dispersive spectrometry was used to locate the heavy metals within the hydrated cement matrix as well as identify various microstructures by determining their elemental composition. These techniques helped locate heavy metals and characterize their effect on temporal hydration.

The following conclusions seem appropriate based upon research conducted at the University of New Hampshire and may or may not relate to all cement stabilized metals:

1. XRD analysis is a viable tool in identifying, tracking, and explaining the effect of heavy metals on the hydration of Portland cement paste.
2. SEM EDAX is essential in qualitatively determining metal distribution within stabilized waste.
3. Equipment limitations and beam spread difficulties precluded the definitive association of the heavy metals with particular hydration products.
4. Binding characteristics of the matrix appear to be significant relative to total porosity or pore size distribution.
5. Heavy metals affect the normal temporal hydration process and was followed by XRD and SEM.

REFERENCES

1. Landreth, R. E. 1986. "Guide to the Disposal of Chemically Stabilized and Solidified Waste." U.S. Environmental Protection Agency, Cincinnati (EPA SW-872).
2. Pojasek, R. B. 1979. "Disposing of Hazardous Chemical Wastes." *Environ. Sci. Technol.* 13(7): 810–811.
3. Poon, C. S., C. J. Peters, and R. Perry. 1983. "Use of Stabilization Processes in the Control of Toxic Wastes." *Effluent Water Treatment J.* 23: 451–459.
4. Poon, C. S., C. J. Peters, R. Perry, and C. P. V. Knight. 1984. "Assessing the Leaching Characteristics of Stabilized Toxic Waste by Use of Thin Layer Chromatography." *Environ. Technol. Lett.* 5: 1–6.
5. Poon, C. S., A. I. Clark, and R. Perry. 1985. "Investigation of the Physical Properties of Cement-Based Fixation Processes for the Disposal of Toxic Wastes." *Public Health Engineer* 13: 108–111.
6. Poon, C. S., C. J. Peters, R. Perry, P. Barnes, and A. P. Baker. 1985. "Mechanisms of Metal Stabilization of Cement Based Fixation Processes." *Sci. Total Environ.* 41: 55–71.
7. Poon, C. S., A. I. Clark, C. J. Peters, and R. Perry. 1985. "Mechanisms of Metal Fixation and Leaching by Cement Based Fixation Processes." *Waste Management Res.* 3: 127–142.
8. Poon, C. S., A. I. Clark, R. Perry, P. Barnes, and A. P. Barker. 1986. "Permeability Study on the Cement Based Solidification Process for the Disposal of Hazardous Wastes." *Cement Concrete Res.* 16: 161–172.
9. Poon, C. S., A. I. Clark, and R. Perry. 1986. "Atomic Structure Analysis of Stabilized Toxic Wastes." *Environ. Technol. Lett.* 7(9): 461–468.
10. Poon, C. S. 1989. "A Critical Review of Evaluation Procedures for Stabilization/Solidification Processes." *Environmental Aspects of Stabilization and Solidification of Hazardous and Radioactive Wastes.* Cote, P. L. and T. M. Gilliam, Eds. American Society for Testing and Materials, Philadelphia (ASTM STP 1033). pp. 114–124.
11. Wiles, C. C. 1987. "A Review of Solidification/Stabilization Technology." *J. Haz. Materials* 14: 5–21.
12. Van der Sloot, H. A., G. J. de Groot, and J. Wijkstra. 1989. "Leaching Characteristics of Construction Materials and Stabilization Products Containing Waste Materials." *Environmental Aspects of Stabilization and Solidification of Hazardous and Radioactive Wastes.* Cote, P. L. and T. M. Gilliam, Eds. American Society for Testing and Materials, Philadelphia (ASTM STP 1033). pp. 125–149.
13. El-Korshi, T. 1988. "Determination of the Deterioration Mechanisms Caused by Seawater Corrosion and Freeze-Thaw Action of Portland Cement Solidified and Stabilized Heavy Metal Wastes." Ph.D. dissertation, University of New Hampshire.

14. Mehta, P. K. 1986. *Concrete: Its Structure, Properties and Materials.* Prentice-Hall, Englewood Cliffs, NJ.
15. Goto, S. and D. Roy. 1981. "Diffusion of Ions through Hardened Cement Pastes." *Cement Concrete Res.* 11: 751–757.
16. Mehta, P. K. and D. Manmohan. 1980. Proc. 7th Int. Congr. on Chemistry of Cements. Paris.
17. Hughes, D. C. 1985. "Sulphate Resistance of OPC, OPC/Fly Ash and SRPC Pastes: Pore Structure and Permeability." *Cement Concrete Res.* 15: 1003–1012.
18. Neuwirth, M., R. Mikula, and P. Hannak. 1989. "Comparative Studies of Metal Containment in Solidified Matrices by Scanning and Transmission Electron Microscopy." *Environmental Aspects of Stabilization and Solidification of Hazardous and Radioactive Wastes.* Cote, P. L. and T. M. Gilliam, Eds. American Society for Testing and Materials, Philadelphia (ASTM STP 1033). pp. 201–213.
19. Birchall, J. D., A. J. Howard, and D. D. Double. 1990. "Some general Considerations pf a Membrane/Osmosis Model for Portland Cement Hydration." *Cement Concrete Res.* 10: 145–155.
20. Birchall, J. D. 1986. "Evidence as to the Formation and Nature of CSH from the Chemistry of Sucrose in OPC Paste." Presented at the Materials Research Society, Symposium M, Boston.
21. Shively, W. 1984. "The Chemistry and Binding Mechanisms Involved with Leaching Test of Heavy Metals Solidified and Stabilized with Portland Cement." Masters thesis, University of New Hampshire, Durham, NH.
22. Nevill, A. M. 1983. *Properties of Concrete.* Halsted Press, New York.
23. Mindess, S. and J. F. Young. 1981. *Concrete.* Prentice-Hall, Englewood Cliffs, NJ.
24. Shively, W., P. Bishop, D. Gress, and T. Brown. 1986. "Leaching Tests of Heavy Metals Stabilized with Portland Cement." *J. Water Pollut. Control Fed.* 58(3): 234–241.
25. Melchinger, K., T. El-Korchi, D. Gress, and P. Bishop. 1987. "Stabilization of Cadmium and Lead in Portland Cement Paste Using a Synthetic Seawater Leachant." *Environ. Progress* 6(2): 99–103.

8 THE CHEMISTRY AND LEACHING MECHANISMS OF HAZARDOUS SUBSTANCES IN CEMENTITIOUS SOLIDIFICATION/STABILIZATION SYSTEMS

D. L. Cocke and M. Y. A. Mollah

8.1. ABSTRACT

Although solidification/stabilization (S/S) technology has been known for a long period of time, the physical and chemical changes that take place as a result of the interaction of the priority metal pollutants (Cr, Pb, Ba, Se, Zn, Ag, Hg, As, and Cd) with cement have not been fully characterized. To understand the important chemistry involved requires the delineation of relevant processes and these will be discussed in the following format:

Nature of surface problem
Characterization methods
Relevant cement chemistry
Nature of active surfaces
    Adsorption phenomena
    Precipitation
Surface reactivity of priority metals
    Carbonation
    Hydration
    Silica polymerization
Leaching mechanisms

0-87371-748-1/93/$0.00+.50
© 1993 by Lewis Publishers

The chemistry and leaching mechanisms of priority metal pollutants in cementitious S/S systems will be discussed in terms of recent results from our laboratory. The nature of the surface species and their surface and solution chemistries will be emphasized. Since characterization of the relevant chemical and physical states is essential, selected characterization techniques are outlined and their application to model doped samples are demonstrated. The results from scanning electron microscopy (SEM), X-ray photoelectron spectroscopy (XPS), Auger electron spectroscopy (AES), ion scattering spectrometry (ISS), mercury intrusion porosimetry (MIP), magic angle spinning-nuclear magnetic resonance (MAS-NMR), Fourier transform infrared (FTIR), and X-ray diffraction (XRD) will be emphasized. It has been found that zinc, cadmium and lead are preferentially deposited on the surface of the cement grains. The surface compounds for Cd and Zn have been identified as mixed hydroxides $CaCd(OH)_4$ and $CaZn_2(OH)_6 \cdot 2H_2O$, respectively. These species apparently result from the Ca aided adsorption of the the normally anionic Cd and Zn species to negative surfaces at high pH. The Pb surface species have yet to be identified but they are likely to be negative species in solution at high pH and will require careful study to determine the nature of the adsorbing entities. Hg has been proven to be present as surface particulate, HgO. Ba has been found to be present as sulfate and carbonate. Cr is incorporated into the C-S-H matrix in some yet to be determined fashion. The results will be summarized in physical chemical concepts. The experimental information should prove useful as a guide for the characterization of S/S systems and the concepts developed should provide for the design of new or improved stabilization and solidification systems. The detailed chemistry is applicable to extend the general chemical knowledge of cement systems.

8.2. INTRODUCTION

S/S[1] became a necessity with the implementation of the Resource Conservation and Recovery Act of 1976 for hazardous substances that are not eliminated by resource recovery, incineration, and/or source reduction. S/S processes based on cement and other pozzolanic materials are being increasingly used for protective environmental measures involving hazardous metals, organic compounds, and inorganic chemicals. The S/S technology uses selected materials like Portland cement, fly ash, pozzolan, lime, etc. for binding the hazardous waste prior to landfilling.

Stabilization is the process whereby hazardous wastes are treated in ways to reduce their toxicity and to prevent dissolution of the toxic components and their release into the environment. The mechanism may be chemical bonding or physical entrapment. Tittlebaum et al.[2] have defined the term chemical fixation, as it refers to the stabilization process, to involve chemical bonding or specific interactions between the waste and the binding agent.

Solidification refers to bringing the waste into a solid state that will allow land filling. It implies no chemical interaction between the waste and the binding agent. Hazardous wastes as aqueous solutions or suspensions are being isolated from the environment by incorporation into cement-based waste containment systems, where the waste is either physically contained or chemically bound or both.

Recent review articles[3,4] and books[5-7] summarize the technology of S/S but there is a need for examination of the chemical aspects involved. This article along with an earlier review[8] is an attempt to delineate the complex chemistry and characterization problems involved in the S/S of metallic hazardous wastes. Hazardous wastes can be organic or inorganic. Organic wastes, which may range from biological organisms to chemicals expelled from industrial processes, are beyond the scope of this review. The inorganic wastes may range from radioactive wastes to incineration residue. In some cases these may be made less hazardous by conversion to other compounds, but toxicity of their elemental forms requires their disposal by concentration and storage.

This article summarizes some of the most recent results from the authors' laboratory and focuses attention on the problem from its chemical basis. It is not to be taken as a comprehensive review. Although we focus on the inorganic non-nuclear wastes that require stabilization and solidification, the chemistry discussed may however also be quite useful for consideration of radioactive wastes. The Environmental Protection Agency, EPA, has identified priority metallic pollutants in this class (i.e., Pb, Cr, Zn, Hg, Cd, Se, Ag, As, and Ba). Since most of these are commonly found as ions (cations or anions), their mutual interactions and their aqueous chemistry must be considered as well as their interactions with the cement components through adsorption, precipitation, or solid solution.

The complexity of the S/S process results from a blending of dynamic cement chemistry with solution equilibria and kinetic processes coupled with the surface and near-surface phenomena. The surface processes are central to many important considerations. Our knowledge of the adsorption, precipitation, and surface processes is far from complete in dynamically reacting systems such as hydrating and setting cement. The mutual effects of the cement and the pollutants need to be determined and incorporated into detailed models of the S/S systems. There is insufficient information on (1) the adsorption behaviors of different metal cations on surfaces of cementitious materials, (2) the chemical and physical changes that take place as a result of adsorption of these metal ions, (3) the effect of cement structure on solidification, (4) the speciation or chemical nature of the metals after S/S, and (5) the leaching behaviors of different metal cations adsorbed on these substances. This requires extensive application of modern surface and bulk analytical techniques. In recent publications[9-19] we have shown how these surface techniques (XPS, AES, ISS, etc.), bulk characterization techniques (XRD, SEM-EDS), and optical spectroscopies (FTIR, solid-state NMR) can be successfully utilized to

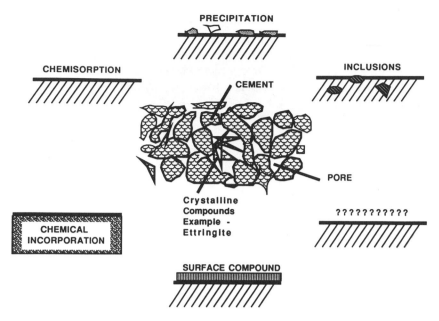

Figure 8-1. Various possibilities for the interaction of hazardous substances with cement.

unveil the chemistry of solidification and lead to better understanding of the leaching mechanisms. Considerable data are presented to demonstrate the application of these methods. *The reader is encouraged to examine carefully the results as they hold the keys to future progress in S/S chemistry.*

The solution, bulk, and interfacial (surface) chemistry of cement is an intimate aspect of the S/S process. However, even the study of the mechanism of cement hydration is still a very active and controversial area. The complexity of the system and the inadequacy of our analytical tools has slowed progress in gaining a complete understanding of cement chemistry. Mix this with the rich chemistry of hazardous wastes and the mutual chemical and physical interactions occurring and one has an extremely challenging problem requiring a highly coordinated characterization effort. Even the most mundane question "where is the hazardous substance located in the cement matrix" is difficult to answer and has not been adequately addressed. Figure 8-1 illustrates the chemical/structural challenge and the several types of interactions that may occur in the solidified system. A waste chemical component may chemisorb, precipitate, form a surface compound to any of several cement component surfaces, form inclusions or be chemically incorporated into the cement structures, or have simultaneous occurrence of several of these situations. It is a major goal of this article to address these issues and to guide the reader in the use of the multitechnique approach to characterization of solidified hazardous wastes.

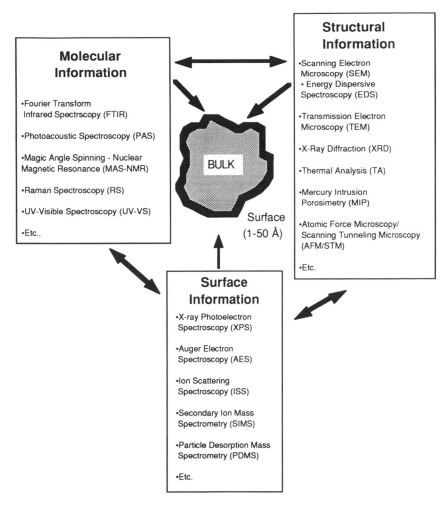

Figure 8-2. Illustration showing the range of characterization techniques needed to study the binding chemistry and leaching mechanisms of stabilized and solidified hazardous substances.

8.3. CHARACTERIZATION APPROACHES

Figure 8-2 illustrates the characterization problems which require a large number of characterization tools to study the nature of the surface and bulk chemistry. Each characterization method with its own set of limitations generally cannot delineate complex heterogeneous systems such as cement, waste chemistry, and the coupled chemistries between them. A group of carefully selected characterization tools is needed. Only a very cursory introduction to the characterization tools will be given. Interested readers are encouraged to examine the reviews and monograms that are referenced.

Physical/Bulk Characterization

Physical characterization involves determining the macroscopic, microscopic, and morphological structure of the material. Optical microscopy can literally identify thousands of compounds by their visual characteristics[20] and has been effectively used on cement systems to identify relatively large features. Electron microscopy by SEM, scanning transmission electron microscopy (STEM), transmission electron microscopy (TEM), and high-resolution transmission electron microscopy (HRTEM)[21] can provide very valuable insight into the microstructure and morphology. Particularly valuable for chemical/physical identification is analytical electron microscopy provided by the SEM and STEM using X-ray analysis.[22] X-ray diffraction[23] provides the best structural look at the solid system. However, it is limited in that it sees effectively only the crystalline components. Since some of the material is expected to be amorphous, XRD cannot be used to describe the complete system. Composition can be supplied by X-ray fluorescence,[24] electron probe techniques such as energy-dispersive analysis of X-rays (EDAX) and wave dispersive analysis of X-rays (WDAX),[22] and UV fluorescence.[25]

Pore structure can be determined by mercury porosimetry.[26] It is particularly valuable when used to provide details of pore structure in the large (hundreds of microns) to submicron dimensions.

Molecular Characterization

Molecular characterization is being provided by FTIR[27] which through vibrational aspects provides insight into molecular structure. During the hydration of cement, the vibrational spectra change with time.

Application of vibrational spectroscopy (IR and Raman) to clinkers and unhydrated cement has been reviewed.[28-30] The vibrational spectral aspects of water and hydroxyl groups in inorganic systems are discussed by Nakamoto.[31] Water in minerals has also been extensively reviewed.[32-38] The vibrational spectroscopy of water and OH in cement has been discussed[39,40] but the complexity of the system has hindered extensive use of the water bands for chemical studies.

Vibrational spectroscopy is being developed to follow the growth of silicate structures in minerals and cement.[28,29,41] XPS[42] and MAS-NMR[43] are being developed for the same purpose. This makes FTIR spectroscopy particularly useful to follow surface chemistry. Unhydrated cements typically show moderate to strong bands at 525 and 925 cm^{-1} from alite. Peaks at 1120 and 1145 cm^{-1} are from S-O stretching vibrations. The bands in the 1650- and 3500-cm^{-1} regions are due to water and OH. The formation of ettringite in the first stages of hydration produces a change in the sulfate adsorption to a singlet centered at 1120 cm^{-1}. Replacement of ettringite by monosulfate causes a return to a doublet at 1100 and 1170 cm^{-1}. Hydration produces a dynamic reaction system. The silicate phases which produce a broad Si-O absorption band at 925 cm^{-1} polymerize to produce a shift to 970 cm^{-1} which eventually obscures the sulfate adsorption band. Hydration also causes changes in the H$_2$O bending band near 1650 cm^{-1} and in the H$_2$O or OH stretching bands in the 3100- to 3700-cm^{-1} region. In OH the stretching modes give a peak at 3640 cm^{-1}, in ettringite one at 3420 cm^{-1} and a weaker one at 3635 cm^{-1} and monosulfate ones at 3100, 3500 (broad), 3540, and 3675 cm^{-1}. The challenge is to unravel the complexity due to partial overlapping of these absorption bands. The water stretching region potentially holds the information needed to identify the nature of the active surface species, the hydroxyl species, and the site specificity for adsorbing species that interact with compounds formed by a metal such as zinc.

Solid-state NMR[44] is providing structural detail around selected nuclei such as Si and Al which are ideal for cement characterization. High-resolution solid-state ^{29}Si NMR techniques have been successfully applied to characterize the hydration products of Portland cement.[45-47] The chemical shift of ^{29}Si nuclei in various silica minerals is dependent on the nature of the X group in Si-O-X units.[48] This chemical shift has been successfully used to characterize the nature of silicate polymers in hydrated cements.[49] The orthosilicate unit SiO$_4^{4-}$ (Q$_o$) in Portland cement reacts with water to give silicate polymers. The chemical shifts of the major components in hydrated cement have been designated as Q$_1$ for terminal SiO$_3^{3-}$, Q$_2$ for internal SiO$_2^{2-}$, Q$_3$ for branching SiO- and Q$_4$ for cross-linking unit Si(OSi)$_4$. Cartledge et al.[49] have reported that the solid-state ^{29}Si NMR technique can be used as an effective tool in elucidating the chemistry of solidified hazardous materials. The polymerization of silicate is being effectively followed by this technique.

Surface Characterization

Surface characterization techniques have only recently been applied to S/S research, but are very relevant since waste components have a high probability of being found on the surface of cement particles. Several techniques have become rather standard methods. These are XPS,[50] AES,[51] ISS,[52] and secondary ion mass spectrometry (SIMS).[53] XPS, by measuring the binding energies of electrons ejected from the surface region of a material by X-rays, provides elemental identification and chemistry (mainly oxidation states) by the shifts in these binding energies. Today, XPS is widely used in surface characterization because it probes the surface core-level electronic states and thus provides qualitative and semiquantitative information concerning the surface. Shifts in the core-level electronic states provide information concerning the chemical state of the surface components as well as knowledge of the bonding interactions between these components.

In recent publications,[53,54-57] we have demonstrated how this surface analysis technique can be utilized for elucidating the binding and chemical environment in clay, clay minerals, and other Si-containing compounds. The position of Si (2p) core electronic states in the silicate minerals has been found to be a diagnostic test for delineating the surface conditions in these substances.[57,58] Carriere et al.[58] have carried out extensive studies of the Si(2p) XPS peaks in Si-containing minerals. Based on the Si(2p) peak positions they have grouped these substances into four categories: (1) elemental silicon at a binding energy (BE) of 99.0 eV, (2) Fe-Si alloys at BE = 100.0 eV, (3) intermediate Si oxidation states or those silicate minerals containing less than four Si-O bonds per Si atom at BE = 102.0 – 103.0 eV, and (4) heavily oxidized Si-containing minerals that contain SiO_4 tetrahedra linked at four corners by -Si-O-Si- bonds at BE = 104.0 eV. AES measures the surface composition by the characteristic kinetic energies of the ejected Auger electrons. Since an electron beam can be focused and scanned across a sample, a lateral resolution of a few hundred angstroms is possible. ISS identifies the outer atomic layers and provides depth profiling information over a few monolayers by measuring the energy of scattered rare gas ions from a surface relative to the impinging ion beam energy. SIMS has the greatest surface sensitivity and provides surface composition by mass analyzing the ions ejected from a surface by a primary ion beam. Many other techniques listed in Figure 8-2 provide complimentary information. In recent publications[9,11-13,15,18,19] we have shown how modern surface analysis techniques (XPS, AES, ISS, etc.), bulk characterization techniques (XRD, SEM/EDS, etc.), and optical molecular spectroscopies (FTIR, Raman, solid-state NMR, etc.) can be successfully utilized to reveal the chemistry and leaching mechanisms of cementitious materials.

The hydration chemistry, the carbonation chemistry and the silicate polymerization process in various S/S systems have been examined and will be reported including Zn, Pb, Cd, Hg, Ba, and Cr doped cement. As an exemplary model the Zn-Cement system will be discussed in detail. Results on characterization and reactivity studies will be correlated with binding models and leaching mechanisms. This paper will place the current and needed research in S/S systems in perspective using recent results correlated with previous studies.

8.4. CEMENT CHEMISTRY

Cement chemistry involves clinker hydration and product dehydration as well as the carbonation of hydroxide components that form. Portland cement clinker whose hydration initiates the chemistry is a mixture of four principal compounds, C_3S (Ca_3SiO_5) 50 to 70%, C_2S (Ca_2SiO_4) 20 to 30%, C_3A $(Ca_3Al_2O_6)$ 5 to 12%, and C_4AF $(Ca_4Al_2Fe_2O_{10})$ 5 to 12%.[59-61] Additional components such as gypsum have been added to alter the hydration chemistry of certain components. C_3A is known to react very rapidly with water followed by the precipitation of calcium aluminate hydrates with considerable evolution of heat.[62] Gypsum slows the rapid setting caused by the C_3A by forming an ettringite coating. Water molecules, hydroxide ions, and chemically bound water are important components of the reacting system. However, C_3S and β-C_2S are also involved in initial setting and provide most of the strength in the first days. C_3S and β-C_2S are metastable phases and have substantial negative free energy of hydration. However, the rates of hydration are slower than for C_3A indicating a substantial activation energy of hydration,[62] which is determined by the nature of the chemical interaction between the water and the surfaces of C_3S and β-C_2S particles. It is generally believed that these have open structures that readily allow water interaction. In addition, CaO is believed to provide favored sites for water attack. Initially on mixing with water, the solution contains dissolved hydroxides and sulfates of calcium, sodium, and potassium. After 10 to 20 h, the sulfate has been almost completely removed from solution by formation of calcium sulfoaluminate. The solution is then mainly that of KOH, NaOH, and saturated $Ca(OH)_2$.

The setting of cement can be divided into four overlapping stages as shown in Figure 8-3:

Stage 1

$3Ca \cdot Al_2O_3 + 3CaSO_4 + 32H_2O \rightarrow 3CaO \cdot Al_2O_3 \cdot 3CaSO_4 \cdot 32H_2O$ (ettringite)

Stage 2

$2(3CaO \cdot SiO_2) + 6H_2O \rightarrow 3CaO \cdot 2SiO_2 \cdot 3H_2O$ (C-S-H) $+ 3Ca(OH)_2$
                          tobermorite-like phase
$2(2CaO \cdot SiO_2) + 4H_2O \rightarrow 3CaO \cdot 2SiO_2 \cdot 3H_2O$ (C-S-H) $+ Ca(OH)_2$

Stage 3

$3CaO \cdot Al_2O_3 + Ca(OH)_2 + 12H_2O \rightarrow 4CaO \cdot Al_2O_3 \cdot 13H_2O$
$4CaO \cdot Al_2O_3 \cdot Fe_2O_3 + 4Ca(OH)_2 + 22H_2O \rightarrow 4CaO \cdot Al_2O_3 \cdot 13H_2O +$
$4CaO \cdot Fe_2O_3 \cdot 13H_2O$

Stage 4

$2(3CaO \cdot Al_2O_3) + 3CaO \cdot Al_2O_3 \cdot 3CaSO_4 \cdot 32H_2O + 4H_2O \rightarrow$
$3(3CaO \cdot Al_2O_3 \cdot CaSO_4 \cdot 12H_2O)$
     monosulfate

The C-S-H material, which is generally amorphous, plays a critical role in cement setting. The mechanism of formation of C-S-H from tricalcium silicate

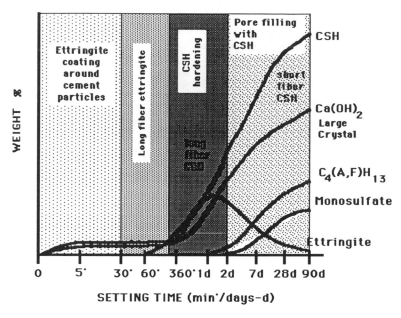

Figure 8-3.   The hydration of Portland cement.

and dicalcium silicate is an active area of cement chemistry that has strong implications in the S/S processes. The hydration and setting chemistry of Portland cement is a very active area of investigation and much remains to be learned. Two models for cement setting have been proposed, the osmotic or gel model[63a,63b] and the crystalline model.[64,65] It appears that the C-S-H gel product is formed at the cement particle surfaces. The membrane formed around the grains allow the inward flow of water and the outward migration of small ions such as Ca^{2+} and OH^-. This results in an excess of $Ca(OH)_2$ on the fluid side of the membrane resulting in precipitation of portlandite while an excess of silicate ions build on the grain side of the membrane producing an osmotic pressure differential on the membrane. The pressure causes the membrane to rupture peridiocally and reforms by extruding concentrated silicate solution. The membrane has been used to explain the retardation of the setting of cement[66,67] in the presence of heavy metal wastes by Double et al.[63,68] and by Birchall et al.[69,70] This provides an important key in studying the hydration mechanism and the interfacial zone of importance in metal waste interaction with cement and may be a dominant feature that relates the solid surface directly to the interfacial chemistry of the solution. A colloidal gel membrane is formed by precipitation at high local levels of supersaturation between calcium in solution and the hydrolyzed silicate-rich surfaces of the cement particles. This protective C-S-H gel coating initially retards the hydration but the membrane is ruptured by osmotic processes allowing a secondary growth of silicate hydrates during the acceleration stage of hydration. The Thomas and

Double model[63] explains the sequence of the hydration reactions and accounts for the mechanism of transport of the silicate material during precipitation of the C-S-H gel. The significance of this is that the surface composition and chemistry of the interface is an important area to be studied.

The crystal model has initial hydration proceeding by nucleation and growth of hexagonal crystals of calcium hydroxides. These eventually fill the spaces between the cement grains. A silicate-rich layer on the surfaces of the cement grains results which slows the interaction of the cement surface with water and retards the release of calcium and silicate ions from the cement. Calcium silicate hydrate grows outward from the cement grains in the form of needle.[64] They eventually come into contact with other needles from other cement grains and produce sheets of tobermorite.

There are probably aspects of both models that have to be considered in detailed mechanisms of hydration and setting. Irrespective of the model chosen, the interaction of the basic products of hydration with carbon dioxide must be considered.

The carbonation of cement has been followed by vibrational spectroscopy.[71-77] It has been found to be a complex process that is dependent on the nature of the cement, its type, porosity, permeability, water to cement ratio,[78,79] as well as the humidity,[80,81] carbon dioxide partial pressure,[82,83] and the nature of the dopant.[62,83-86] Carbonation of the C-S-H converts the OH^- and the Ca^{2+} to calcium carbonate and results in the apparent formation of a highly polymerized silica gel which is acid stable and maintains the same morphology as the original hydrate.[74,76] In addition, the calcium hydroxide can react directly with the carbon dioxide. These reactions are

$$C - S - H + CO_2 \rightarrow CaCO_3 + silica + H_2O$$

$$Ca(OH)_2 + CO_2 \rightarrow CaCO_3 + H_2O$$

which have been discussed.[72,78,81] The first reaction appears to be pseudomorphic with no change in morphology.[74,77,87] The second reaction should result in a volume expansion and could result in structural problems[88] if it occurs in the bulk rather than at the surface. There is disagreement as to which of the above reactions dominate.[78,81,89] Carbonation of hydrated cement needs thorough investigation since it can neutralize the alkalinity of cement paste which could affect the solidification and stabilization of wastes.

The extremely rich ion chemistry must be considered in the system. The metal ions can exchange with the ions in the C-S-H. They can precipitate with the anions that are a part of the clinker or gypsum. They can become included in the C-S-H. They can chemisorb at ionic or charge sites at the surfaces. Chemisorption phenomena at the liquid solid interface will depend on the surface structure of the C-H-S in a very basic environment.

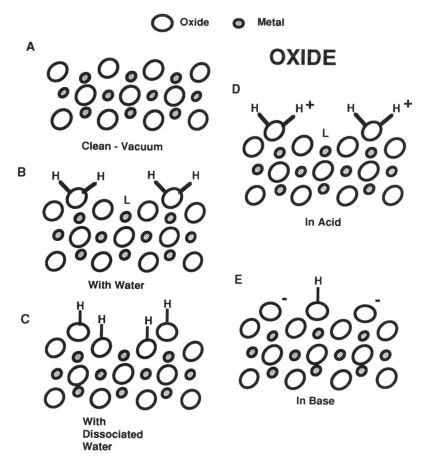

Figure 8-4.   Cross-sectional views of the surface regions of an oxide under various conditions: (A) oxide surface formed in vacuum, (B) oxide surface with molecularly adsorbed water, (C) oxide surface with dissociatively adsorbed water, (D) oxide surface in acid solution, (E) oxide surface in basic solution. L is a Lewis acid site.

8.5. CHEMICAL SURFACE STRUCTURE

The surface structure of the hydration products and the C-S-H in a basic solution as deprotonated hydroxlated surface is depicted in Figure 8-4E. The interactions of metal ions with hydrous oxide surfaces have been extensively studied. Quantitative predictions of the extent of the pH-dependent reversible adsorption have been considered.[90-96]

Adsorption

The surfaces an adsorbate sees under various conditions are illustrated in Figure 8-4. Figure 8-4A shows an oxide surface that has been created in vacuum. Exposure of this surface to water results in the adsorption of water molecules (Figure 8-4B). These can dissociatively chemisorb to produce surface hydroxyls that have both acidic and basic properties (Figure 8-4C). In acid solution such as might be experienced in leaching, the basic hydroxyls are removed from the surface-producing Lewis acid sites that may have a full-positive or partial-positive charge. These are marked L in Figure 8-4D. The acidic hydroxyls can add additional protons producing Bronsted acid sites with positive charges. In acidic solutions, the surface can pick up positive charge which restricts cation adsorption and in leaching can cause cation ejection. In basic solutions, as commonly found in cement, the surface hydroxyls that can give up protons are deprotonated producing negative oxide sites (Lewis base sites). The basic hydroxyls remain on the surface but a net negative charge will develop that restricts anion adsorption, causes anion ejection, and promotes cation adsorption.

Typical anion and cation adsorption curves as a function of pH are shown in Figure 8-5.

The surface complexation model[90,91] treats specific adsorption as the interaction of the metal ion with the oxide surface in analogy to the complexation of metal ions by ligands in solution. Surface hydroxl groups have the coordination properties of oxygen donor ligands. Protons and metal cations compete for the oxygen coordination sites on the surface. The major reaction with the negative silicate surface is competitive adsorption between the cations such as H^+, Na^+, K^+, and Ca^{2+} and since at the high pH values protons are in extremely low concentration the latter ions are competing for surface sites with the hazardous ions. In principle a set of equilibrium constants can be written that permits estimation of the surface speciation; however, the complexity and lack of information on the composition and its rapidly changing nature in cement makes this approach difficult. The complexation of metal ions by oxide surfaces occurs over a very narrow pH range and at high pH values the adsorption of metal ions is 100%. In addition to adsorption, the direct precipitation of oxides, oxyhydroxides, and hydroxides will occur at the surface along with the expected adsorption reactions.

Corey[97] has discussed adsorption vs. precipitation and summarized the processes:

1. Crystal growth occurs if the adsorbate is a component of the cement substrate.
2. Crystal growth and/or diffusion into the solid phase occurs if the adsorbate is not a component of the adsorbent but can form a solid solution.

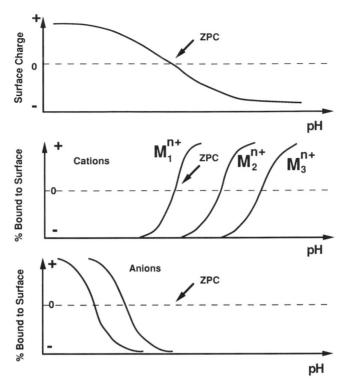

Figure 8-5.   The pH response of cation and anion adsorption on hydrated oxide surfaces.

3. Formation of a stable surface compound can take place if the adsorbate is not capable of forming a three-dimensional solid solution with the adsorbent.
4. Stabilization of metastable polynuclear ions occurs by adsorption onto oppositely charged surfaces of the adsorbent.
5. Heterogeneous nucleation of a new solid phase may involve a new phase composed of the adsorbate and a component from the solution (hydroxides, carbonates, etc.).
6. Heterogeneous nucleation of a new solid phase may occur where the new phase is composed of the adsorbate and a component of the adsorbent resulting in dissolution of the adsorbent and redeposition as a different compound.

All these processes are potentially active in the cement stabilization and solidification process.

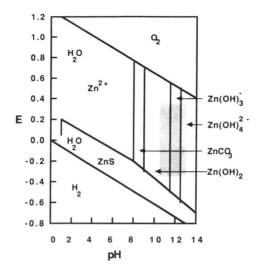

Figure 8-6.   **Stability of some zinc compounds in relation to E (volt) and pH.**

Aqueous Chemistry of Metals

The prediction of the speciation of the metals in the cement is dependent on the solution chemistry of the metal ion in very basic solutions containing sulfate and eventually some carbonate. Both the pH and the E_H must be considered. Zinc is a good example. As shown in Figure 8-6, zinc is expected to form hydroxides in solution in the pH range (>8) and E_H range (shaded area) expected in cement. Zinc hydroxide, a typical ampholyte, functions both as an acid and a base. The equilibria are

$$Zn^{2+} + 2OH^- = Zn(OH)_2 = 2H^+ + ZnO_2^{2-}$$

$$\uparrow\downarrow$$

Acid	ZnO + H$_2$O	Base
pH = 3–5		pH = 11–13

Considering water in the equilibria, the hydroxy-complexes $Zn(OH)_4^{2-}$ and $Zn(OH)_3^-$ are present in strongly basic solution. Their anionic properties preclude their adsorption to the negative surface of the C-S-H at high pH. How adsorption is made possible is a key point to understanding the S/S of zinc. The E_H-pH diagrams of most heavy metals are comparable to that of zinc in that

Me^{2+} is stable at pH less than 7 to 8 and with increasing pH, the carbonate becomes stable which is followed by several hydroxide species.

A lead ion forms $Pb(OH)_2$ in a dilute hydroxide solution and the plumbite ion, PbO_2^{2-}, in a concentrated hydroxide solution. Given the anionic nature of the plumbite, it is also unlikely to adsorb to the negative oxygen sites on the silicate. This is supported by the observation that anions adsorb best at low pH while cations adsorb best at high pH.[98] This is illustrated in Figure 8-6. It can be argued that the lead may be deposited as a precipitate. The simultaneous presence of sulfate and lead ions is expected to produce lead sulfate which is similar to barium sulfate but because of its greater density it is deposited more readily. It is not clear how a precipitate can quicky cover the hydrating clinker particles with a membrane precipitate. One then may be forced to look at adsorption once again and ask if there are any cationic species that are generated in the process that can adsorb on the negative surface. At high pH values Pb(II) may form cluster ions containing up to six Pb atoms such as $[Pb_6O(OH)_6]^{4+}$. In the cement environment, these hydroxy-cations may adsorb to the silicate surface and/or precipitate as sulfates forming a membrane that slows the hydration process.

Barium is expected to produce barium sulfate and some carbonate in cement by precipitation.

Hg^{2+} precipitates as red or yellow (depending on particle size) HgO in alkali hydroxide solutions and is expected in the cement environment.

The most stable aqueous solutions of Cr are the Cr(III) and Cr(VI) systems. The Cr^{3+} ion in aqueous medium is likely to be present as hydrated ion, $[Cr(H_2O)_6]^{3+}$. The hexaquo ion is highly acidic ($pK_a = 4$) and may undergo deprotonation reaction leading to the formation of hydroxy-bridged polymeric species of high molecular weight. In an intermediate alkaline medium $Cr(OH)_3$ may also be formed, which dissolves at high pH to form $[Cr(OH)_6]^{3-}$ and $[Cr(OH)_4]^-$ ionic species. Above pH = 8 the $Cr(OH)_4^-$ may exist in appreciable concentration.[99] These anions are unlikely to be adsorbed by the negative silicate surfaces, since cement paste is highly alkaline (pH = 11 to 13). At present the nature of the Cr species that interact with cement is not known. Complex precipitation or the formation of solid solutions may need to be considered here.

8.6. PHYSICAL STRUCTURE

C-S-H Structure

The gel is generally regarded as a three-dimensional assemblages of C-S-H layers that form subparallel groups of a few layers thickness. Calcium hydroxide, CH, is present as isolated masses and as part of tobermorite where the CH is layered between silicate layers. The physical structure changes with time.

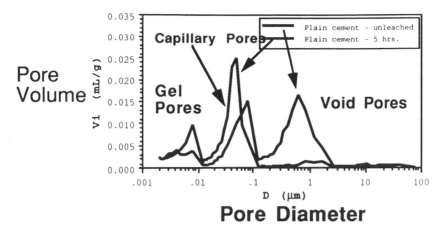

Figure 8-7.   MIP determination of the pore size distribution in unleached and leached cements.

Pore Structure

Hardened cement is a very porous material with a trimodal pore distribution as determined by MIP. The MIP method is based on the behavior of nonwetting liquids in capillaries. The interfacial tension opposes the entrance of the liquid into the pore which can be overcome by external pressure. For a cylindrical pore the force opposing the entrance is acting along the circumference and equals $-2\pi r\sigma \cos \theta$. The opposing external force is $\pi r2P$. At equilibrium the two forces equal and the radius is given by: $r = -(2\sigma \cos \theta)/P$ which reduces to $r = 7500/P$ using the surface tension for Hg to be 0.48 N/m and a average wetting angle of 140°. Typical MIP data for plain cement leached and unleached are shown in Figure 8-7. The smallest pores, gel pores, are approximately 0.01 μ in diameter. These are likely to be associated with inner layer spacing. The next larger pores averaging about 0.1 μ in diameter are due to the original water-filled space in porous cement gel and are called capillary pores. The volume of these pores is directly proportional to the water-cement ratio. The largest pores are formed by air voids. The pore system has continuity as demonstrated by permeability.

8.7. RECENT RESULTS AND DISCUSSION

Recent results from surface analysis techniques and physical structure techniques have begun to provide insight into the chemical binding and interactions between hazardous metal substances and Portland cement. The essence of this work can be illustrated by considering several key questions and the techniques and concepts used to address them:

1.   Where are the hazardous metals located? (XPS, ISS, AES)
2.   What is the chemical nature (speciation) of the metals associated with the cement? (FTIR, Raman, MAS-NMR)
3.   What are the mutual influences of the metals and cement on each other? (SEM, EDS, hydration, and carbonation chemistries)
4.   What surface, solution, and solid-state chemistry control the S/S process? (Speciation and mechanisms of formation)

The XPS and FTIR techniques have been used to examine Portland cement doped with Pb, Cr,[9,11,15] Hg,[12] Zn, Cd,[18,19,100] and Ba.[13] The strong differences in Pb and Cr have been investigated further using mercury porosimetry and modeling.[101] In addition the leaching of these samples was studied by similar approaches.[102] Location of the metals is best illustrated by comparing Pb and Cr.[9,11,15] We have made a comparative study using XPS, ISS, EDS, and SEM. A significant finding of this research is the preferential deposition of lead at the surface of the cementitious material. This observation combined with the fact that lead salts, such as carbonates, silicates, etc. are extremely insoluble and have low surface energies, serve to give credence to the proposed surface sites for Pb ions[60,62] and strongly supports the coating of clinker particles discussed above by Pb species. Since Pb ions drastically retard setting,[67] it must be intimately involved in blocking the hydration mechanisms. Since the hydration of the C_3A and β-C_2S occurs at dramatically different rates and by different mechanisms the Pb ions apparently interfere with both. This would support Pb species creating a barrier to water interaction. Figure 8-12 shows a comparison of the XPS signals from 10 wt% Pb and Cr in Portland cement showing the surface location of Pb.

The metal systems will be discussed individually and cross correlations made where necessary. The reader not interested in the detailed chemistry is advised to go directly to the Zn-Portland Cement Section and the Conclusion Section.

Mercury-Portland Cement (Precipitation Controlled)

The chemical precipitate nature of certain metals in cement is illustrated by the Hg[12] and Ba[13] cases. The chemical nature of the metal species has been delineated. The XPS spectrum of the Hg-doped sample is shown in Figure 8-8.

Here the Hg oxidation state is identified as 2+. The Hg is present as highly dispersed HgO. The color changes from yellow to red on heating the sample at about 40°C. This is commonly observed for HgO in small crystallite form.

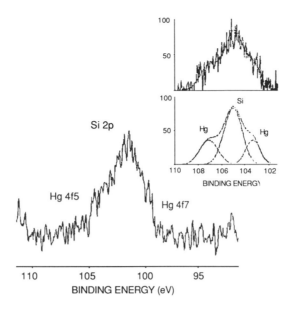

Figure 8-8.   **XPS spectra of Hg-doped cement showing the Si 2p region flanked by the Hg doublet for the $4f_{7/2}$ and $4f_{5/2}$. The insert shows the decomposed spectrum.**

Barium-Portland Cement (Precipitation Controlled)

The presence of $BaSO_4$ is supported by FTIR studies as well.[9,13] Barium-doped Portland cement has been investigated by XPS and FTIR techniques. Figure 8-9 shows the FTIR spectrum of barium sulfate.

The presence of potassium on the surface, in addition to the carbonate, is illustrated in Figure 8-10. It shows that the Ba produces additional carbonate at the surface.

Figure 8-11 clearly shows that two types of sulfate are present at the surface: one is associated with the Ba and the other associated with the sulfoaluminate. This agrees with the FTIR results insofar as both techniques can identify $BaSO_4$. XPS has identified Ba being present as $BaSO_4$ and $BaCO_3$.

The presence of HgO and $BaSO_4/BaCO_3$ show clearly that the normal chemical forces are working and that Hg and Ba are deposited as their very insoluble precipitates. The barium carbonate comes from the reaction of carbon dioxide with the excess (exceeding the sulfate concentration) Ba that is present as the hydroxide.

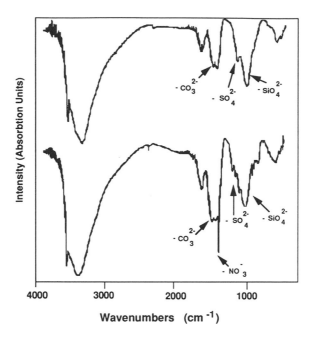

Figure 8-9.   FTIR spectra of (top) undoped Portland cement and (bottom) Ba-doped Portland cement.

Chromium-Portland Cement (Bulk Incorporation)

We have found that chromium is present mainly in the bulk of the medium,[11] as XPS shows a very low concentration of chromium on the surface of Cr-doped cement (see Figure 8-12). The striking difference between the surface concentration of Cr- and Pb-doped samples is shown in Figure 8-12. The very small signal-to-noise for Cr suggests that the Cr is located below the 15- or so angstrom depth probed by the XPS technique.

Speciation of Cr from FTIR

Additional support for the Cr incorporation model comes from FTIR (Figure 8-13) and EDS (Figure 8-14) examinations of the Cr-doped samples before and after leaching. The appearance of poorly resolved bands (Figure 8-13) at around 1048 to 1001 cm^{-1} and the overlapping of v_4 and v_2 SiO_4^{4-} bands suggests that either Cr^{3+} ions have been incorporated with the C-S-H or the polymerization of the silicates have been affected by the reactions and/or adsorption of chromium on the components of OPC. The peaks 1387 cm^{-1} (s, s) and 1353 cm^{-1} (sh,w) are due to v_3 NO_3^- and v_1 NO_3^- vibrational bands from doping solutions.

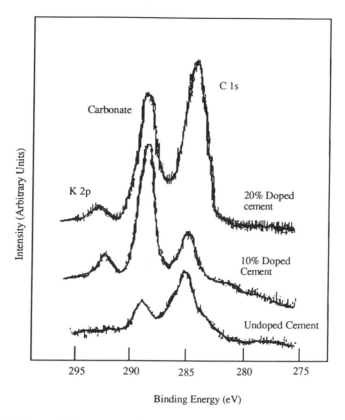

Figure 8-10.   XPS spectra of barium-doped cement (upper two spectra) and undoped cement (lower spectrum).

The FTIR spectra of the leached Cr-OPC sample is given in Figure 8-13c. The reduced intensities of the two carbonate bands centered at 1560 and 1455 cm^{-1} are attributed to the dissolution of carbonates in acidic leaching solution. A strong and broad band due to v_3 SiO_4^{4-} appear at 1060 cm^{-1} and the bands due to v_2 and v_4 vibrations appearing between 525 and 450 cm^{-1} have also been resolved. Close examination of the silicate bands between the unleached (Figure 8-13a) and leached (Figure 8-13c) Cr-OPC samples reveal some critical differences invoked by leaching procedures. The SiO_4^{4-} band in silica phase has now been shifted by about 110 wave numbers in the leached sample and the relative intensities of the v_2 and v_4 vibrations have also been significantly changed. It can be concluded that some chromium has been leached out into the solution, thus exposing the C-S-H to the attack by acidic leaching solution which leads to enhanced degree of polymerization.

The EDS spectra of Cr-doped Portland cement before and after leaching are shown in Figure 8-14. The Ca/Si ratio is in the leached sample rather than the unleached sample. The concentrations of chromium (Figure 8-14a and b) in

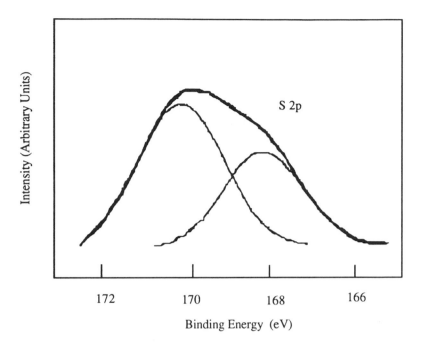

Figure 8-11.   XPS spectra of sulfur 2p in Ba-doped cement. The two peaks are for sulfate assiociation with Ba (168 eV) and aluminosilicate (170 eV).

both samples remain fairly unchanged indicating that the leaching process did not have any effect on the chromium present in the bulk of the solidified matrix. Since EDS is not a surface technique[50,51] and has a penetration greater than 1 µm, it is evident that chromium is dispersed below the surface (50 Å) of the medium. XPS examination of the unleached Cr-OPC sample confirmed[11] that chromium is present in its Cr^{3+} oxidation state and the binding energy was found to be consistent with carbonate, silicate, and hydroxides only and not of oxides.

The silicate regions hold much of the information about the mutual interactions of the dopant metals and the silicate moiety present in OPC. The effect of carbonation on the polymerization of the orthosilicate units present in OPC and the influence of metal pollutants thereon is an important area of research and has broad implications on the S/S systems. We shall, therefore, concentrate our discussion on these two aspects only. In a previous report[18] we have discussed the mutual interactions of metal dopants and the water.

The FTIR results indicate some association of chromium with C-S-H, although it is rather difficult to predict the actual nature of chemical association from the present results alone. The degree of hydration in OPC will affect the intensities of the bending modes v_2 and v_4 and it is also generally expected that

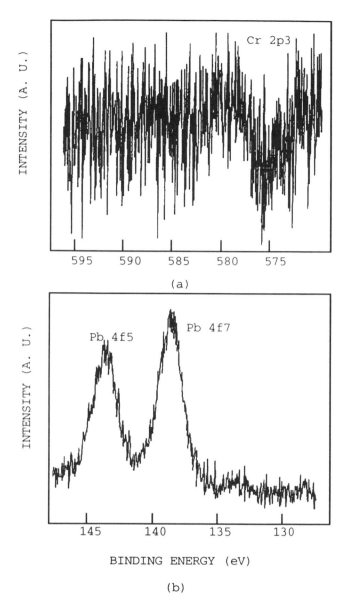

**Figure 8-12.   XPS spectra of chromium- and lead-doped cement sur-
faces.**

the intensities of these vibrations will decrease with increased polymerization.
If one compares the FTIR spectra of Cr-OPC (Figure 8-13a and b) unleached
samples and Cr-OPC (Figure 8-13c) leached sample, it can be clearly seen that
the two bending bands centered between 450 and 525 cm⁻¹ in OPC after

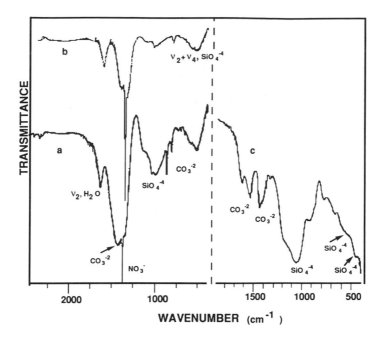

Figure 8-13.   FTIR spectra of (a) Cr-OPC, air cured, (b) Cr-OPC, argon cured, and (c) Cr-OPC, leached.

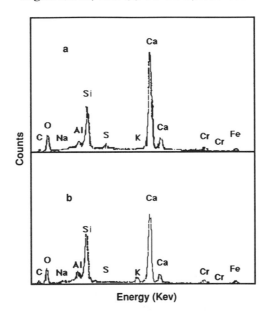

Figure 8-14.   EDS spectra of chromium-doped Portland cement: (a) unleached sample and (b) leached sample.

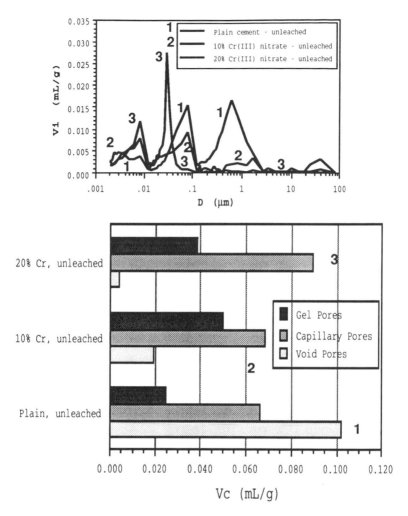

Figure 8-15. MIP determination of the pore size distribution in Cr-doped Portland cement.

leaching appear as a single peak centered at 520 cm^{-1} in Cr-OPC. The major band, v_3, now shows some definitive structure with two closely spaced peaks between 1000 and 1050 cm^{-1}. The leaching process thus seems to have reversed the situation with v_3 now showing as an intense and broad peak centered at around 1060 cm^{-1} and the v_2 and v_4 have again been resolved. The v_2 peak is now centered at 460 cm^{-1}, while the v_4 is barely visible as a shoulder. These vibrational changes are indicative of increased polymerization, as well as chemical association of chromium with C-S-H. Similar chemical reactions of some heavy metals with the synthetic hydrous calcium silicates such as tobermorite ($5CaO \cdot 6SiO_2 \cdot 5H_2O$), xonotlite ($6CaO \cdot 6SiO_2 \cdot H_{22}O$), and wollastonite ($CaO \cdot SiO_2$) have been reported by Komarneni et al.[103a,103b]

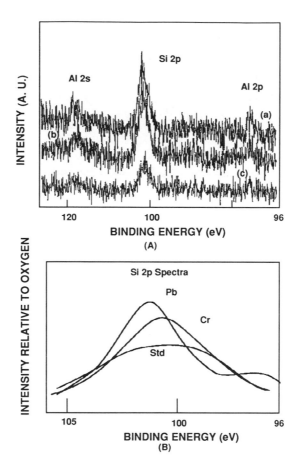

Figure 8-16.   XPS spectra: (A) the Al 2p, Al 2s and Si 2p lines for (a) Pb-doped, (b) Cr-doped, and (c) undoped standard ement sample; (B) Si 2p spectra standardized to oxygen for Cr, Pb, and standard cement.

Hg porosimetry data (Figure 8-15) agree with the above results for chromium-doped systems. The chromium addition affects the pore distribution significantly. It causes a loss of void pores, produces a wider variety of pore diameters in the small (0.01 μm) diameter region, and causes a clear and large shift of capillary pores to smaller size. Cr addition affects the small pore formation mechanisms more than Pb addition, since chromium probably gets more involved in the formation of the silicate structures. Pb addition reduces the number of large void pores and increases the relative number of pores in the capilliary pore region.

Lead-Portland Cement (Surface Deposition)

The location of the Pb in Portland cement is discussed here. Leaching results are discussed and compared to zinc in the zinc section. From the XPS spectra in Figure 8-16, it is clear that the Pb is located on the outer surface of the cement particles while Cr in Cr-doped samples is below the surface. Since XPS and ISS show that the lead is located mainly at the surface, it would suggest that a lead compound is coating the clinker components and is responsible for the initial retarding of cement hydration. This observation combined with the fact that lead salts such as carbonates, sulfates, etc. are extremely insoluble and have low surface energies[60] suggests that lead ions selectively coat the outer portions of the cement components and are not selective about which components are coated.

The shift in Si 2p binding energy for the Pb-doped system as compared to that for the Cr and standard undoped cement is shown in Figure 8-16 and suggests that Pb is associated with the C_3S and β-C_2S components.

The peak position for the Si 2p indicates enhanced polymerization over that of the Cr doped and the standard. The situation is different for the chromium-doped samples in which no chromium is detected on the surface. Apparently, size and charge differences allow Cr to be incorporated throughout the solid material during hydration.

XPS also shows that lead and chromium are present in their original oxidation states, Pb^{2+} and Cr^{3+}, respectively. Their binding energies are consistent with silicate, carbonate, or hydroxide compounds and not with their oxides.

The appearance of potassium at the surface of the doped samples and its absence from the surface of the standard Figure 8-17 suggests that the doping is causing a redistribution of the K^+.

Apparently the K^+ is moving into solution to charge compensate the NO_3^- anions as the Pb^{2+} and Cr^{3+} are deposited in the solid state. Potassium is then deposited at the surface as the liquid is lost to evaporation and chemical reaction with clinker components. In addition the enhanced presence of carbonate in the doped samples is suggested.

The overall Ca/Si ratio of the Pb-doped cement obtained by EDS was slightly lower than that of the standard. These reduced Ca/Si ratios support the findings of Thomas et al.[63] and our XPS results showing increased Si at the surface of the Pb-doped sample. XPS results shown in Figure 8-16 also show that lead enhances the degree of polymerization of the silicate at the surface. FTIR shows the inner silicate to be less polymerized than for the undoped cement.

Zinc-Doped Portland Cement (Surface Deposition)

Zinc-doped Portland cement samples (unleached and leached) have been investigated by XPS, AES, EDS, and SEM techniques. The results indicate that

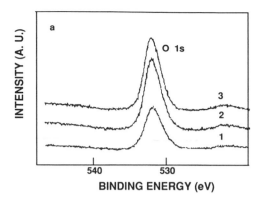

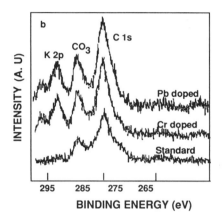

Figure 8-17.   XPS spectra showing the relative amounts of surface carbon, potassium, and carbonate.

Zn is mainly on the outer surface and is effectively removed during acid leaching.

Location of Pb and Zn and Leaching

XPS O1s spectra have been used to differentiate between bridging (-Si-O-Si-) and nonbridging (-Si-O-X-, where X = Na, Al, Fe, etc.) oxygen atoms in minerals and ceramics.[58] The binding energies of the O 1s peaks in both lead and zinc unleached and leached samples occur at about 531.5 and 532.5 eV, respectively. Therefore, a shift (see Figure 8-18) in the binding energy of the O 1s peak of about 1 eV occurs. An O 1s binding energy of approximately 531.0 to 531.5 eV is considered to be characteristic of the O^{2-} oxidation state for hydroxyl, carbonate, and silicate anion species in a cement matrix.[11,50,58] A

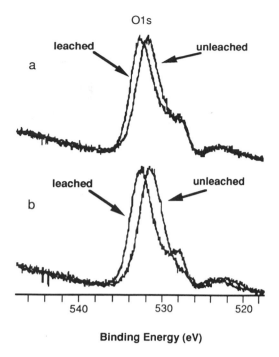

O1s

Figure 8-18.   **XPS spectra of O lines for (a) zinc-doped cement and (b) lead-doped cement.**

shift to higher binding energies due to leaching is indicative of the oxygen reacting with the silicon to form higher polymers. The Si 2p signal has been used to elucidate the structural nature of silicate minerals.[54,58] Carriere et al.[58] have shown that the diagnostic Si 2p XPS peaks for a heavily oxidized Si-containing mineral occurs at about 104 eV, while an intermediate Si-oxidation state for these silicate minerals occurs at around 102 to 103 eV. XPS spectra of Si 2p lines for leached and unleached samples are shown in Figure 8-19a for zinc-doped cement and in Figure 8-19b for the lead-doped sample. The peaks appear at 101.2 and 102.7 eV for the unleached and leached lead-doped samples and the corresponding peaks in the zinc-doped samples appear at 101.0 and 102.6 eV, respectively. The XPS peaks in Si 2p have thus been shifted significantly by about 1.5 eV due to leaching and is therefore indicative of an increase in the number of Si-O bonds per silicon atom in agreement with the O 1s results. Thus, leaching leads to enhanced polymerization of silica present in cement for both lead and zinc samples.

The $Ca(2p_{3/2})$ and $Ca(2p_{1/2})$ peaks in the lead-doped samples appear at 347.1, 350.6, 348.1, and 351.6 eV in the unleached and leached samples, respectively, while the corresponding peaks in the zinc-doped spectra appear at 347.0, 350.5, and 347.4, 350.9 eV, respectively. The binding energy difference between the $Ca(2p_{3/2})$ and $Ca(2p_{1/2})$ peaks is about 3.5 eV, which is

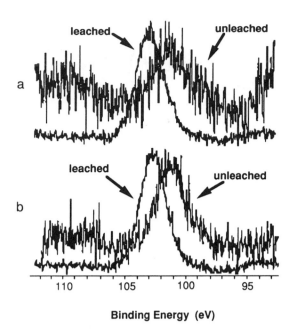

Binding Energy (eV)

Figure 8-19.   **XPS Si 2p spectra: (a) zinc-doped cement and (b) lead-doped cement showing the effects of leaching.**

typical of calcium hydroxide, sulfate, silicate and carbonate.[12,100] The binding energies of Ca ($2p_{3/2}$) peaks in the unleached lead (347.1 eV) and zinc (347.0 eV) doped samples are similar and approximately equal to the same peak in calcium hydroxide which appears at 346.7 eV (Figures 8-20 and 8-21). It, thus, appears that surface calcium in hydrated zinc- and lead-doped cement is present mostly as calcium hydroxide. This is in agreement with the values reported by Suguma[104] from XPS studies of Ca-based hydrates formed by the interaction of $Ca(OH)_2$ and the chemical constitutents present on the surface of ceramic microspheres. The author has reported that the binding energy due to the $2p_{3/2}$ peak appears at 346.7 for $Ca(OH)_2$ and 347.6 and 348.4 eV due to the association of calcium in $CaO\text{-}SiO_2\text{-}H_2O$ and $CaO\text{-}Al_2O_3\text{-}SiO_2\text{-}H_2O$, respectively. However, the binding energies due to the $Ca(2p_{3/2})$ peaks appear at higher positions in the lead- (Figure 8-20) and zinc-doped (Figure 8-21) samples, although the difference between the leached and unleached samples is much more significant for the lead sample (1.0 eV) than the zinc-doped sample (0.4 eV). The reasons for this are not yet known.

$Pb(4f_{7/2})$ peaks in the lead-doped samples appear at 138.4 eV for the unleached sample and 138.8 eV for the leached sample. The relatively small shift in the binding energy indicates that the chemical state of the lead has not changed significantly upon leaching. The binding energy of the $Zn(2p_{3/2})$ peak

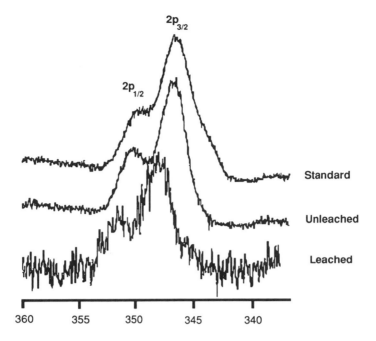

Figure 8-20. XPS spectra of Ca $2p_{1/2}$ and Ca $2p_{3/2}$ lines for lead-doped cement and standard calcium hydroxide.

appears at 1021.5 eV in the unleached sample. However, no zinc was found in the leached sample indicating that most of the surface zinc has been dissolved during the leaching tests.

The lead- and zinc-doped samples were also analyzed using scanning electron microscopy (SEM) and energy dispersive spectroscopy (EDS) for qualitative and morphological information of the leached and unleached surfaces.

The SEM micrographs of the unleached and leached front show two distinctive facets of unleached and leached areas. The characteristic morphology of the phases is also quite visible. The inner continuous matrix is similar to the unleached side. The leached sides consist of a "honeycomb" type of structure made up of flakey bright substances. The morphology of the outer leached side may be compared with a cross-linked, poorly crystalline structure and is similar to ettringite commonly present in hydrated cement. The unleached sides of both lead- and zinc-doped cement consist of a continuous dark area surrounded by bright flakey material. The zinc-doped cement has more amorphous characteristics than the lead-doped cement, while the lead-doped cement shows more needle-shaped structures. The leached sides consist of loosely bound flakey materials with cracks and needle-shaped substances spread over the

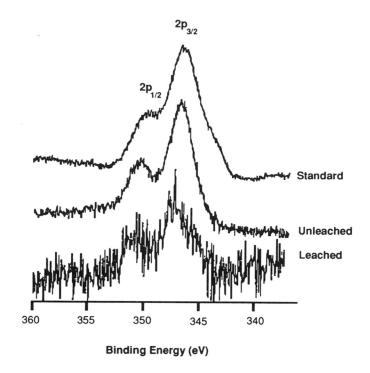

Figure 8-21.   XPS spectra of Ca $2p_{1/2}$ and Ca $2p_{3/2}$ lines for the zinc-doped cement and standard calcium hydroxide.

side. The morphology of the zinc-doped leached sample is fairly similar to the lead-doped leached sample.

The corresponding EDS spectra of the unleached and leached samples are shown in Figure 8-22. The following elements have been identified from the lead-doped sample: oxygen, calcium, silicon, aluminum, lead and potassium. It can be seen (Figure 8-22a and b) that the concentration of silicon and calcium have been drastically changed due to leaching. The calcium to silicon (Ca/Si) ratio has been considerably reduced in the leached sample due to the dissolution of $Ca(OH)_2$. The concentration of aluminum and potassium has increased slightly as a result of leaching, while the concentration of lead has been decreased.

The elements identified from EDS spectra of the zinc-doped sample include oxygen, calcium, silicon, aluminum, and potassium. No zinc or sulfur was detected in the leached sample. It is quite clear (Figure 8-22c and d) that the calcium-to-silicon ratio (Ca/Si) is considerably lower in the leached sample as in the case of the lead-doped cement. These results indicate the presence of higher concentrations of silicon in the near-surface regions of the leached sample. The EDS results are compatible with the XPS results presented above.

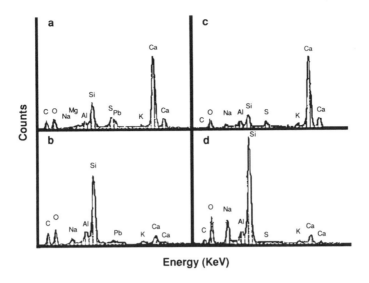

Energy (KeV)

Figure 8-22.   EDS spectra of lead- (a and b) and zinc- (c and d) doped cement: (a and c) unleached sample, (b and d) leached sample. Note the virtual absence of potassium.

Speciation of Zn from FTIR

The addition of zinc and lead to cement has been previously studied. Tashiro et al.[105-108] examined the effects of ZnO and PbO on C-S-H, tobermorite, and related phases. They found that at 180°C willemite (Zn_2SiO_4), hardystonite ($Ca_2ZnSi_2O_7$), and a special phase ($Ca_2Pb_3Si_3O_{11}$) formed under hydrothermal conditions. In water at ambient temperatures, zinc hydroxides are formed as a function of pH as seen in Figure 8-6. Without cation participation these negative species would not be expected to adsorb to the negative C-S-H surfaces found at the high pH in cement.

In the studies of Arliguie et al.,[109,110] $CaZn_2(OH)_6 \cdot 2H_2O$ has been proposed to form by

$$2Zn(OH)_{3^-} + Ca^{2+} + 2H_2O \rightarrow CaZn_2(OH)_6 \cdot 2H_2O$$

$$2Zn(OH)_2 + Ca^{2+} + 2OH^- + 2H_2O \rightarrow CaZn_2(OH)_6 \cdot 2H_2O$$

and, as seen in Figure 8-6, this is pH dependent.

Comparison of the FTIR spectra for dry clinker and hydrated cement shown in Figures 8-23a and b set the stage for further discussion and provide insight

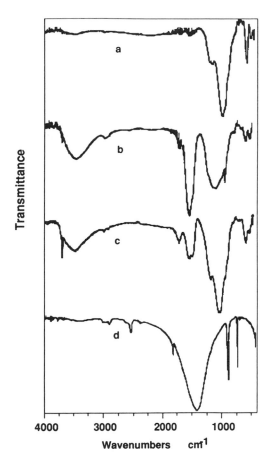

Figure 8-23.   FTIR spectra of Portland cement type v: (a) dry clinkers, (b) hydrated cement cured in air, (c) hydrated cement cured in Ar, and (d) calcium carbonate.

into the reaction between CO_2 and the C-S-H or the CH. The sharp band in the water stretching region at 3644 cm^{-1} is due to the OH stretch of the $Ca(OH)_2$. Note its virtual absence on the dried clinker but its prominence in the hydrated sample. The sample cured under air has a much smaller peak than that cured under argon (Figure 8-23c). The prominence of the carbonate band in the former shows that the CH reacts with the CO_2 from air to produce carbonate. This provides evidence that the CO_2 is reacting with the CH as shown above. However, the decrease in intensity of the v_4 525 cm^{-1} band due to SiO_4^{2-} and the shift of the v_3 SiO_4^{2-} in Figure 8-23b as compared to Figure 8-23c suggests the enhanced polymerization of the silicate due to the carbonate reaction with C-S-H. This shift is seen to obscure the triply degenerated bands at 1101 to 1157 cm^{-1} due to the sulfate as discussed previously. Thus, both reactions

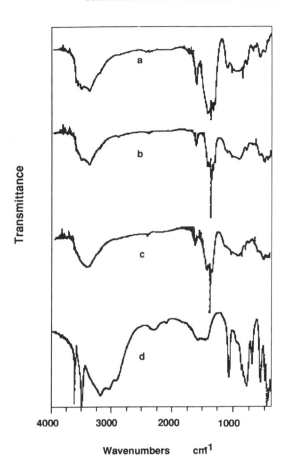

Figure 8-24.   FTIR spectra of Zn-doped cement (a) cured in air, (b) cured in Ar, (c) heated at 150°C for 2 h, and (d) CaZn$_2$(OH)$_6$·2H$_2$O.

appear to be competitive, or the reaction of carbon dioxide with hydroxide provides water that enhances the polymerization of the silicate. It is to be noted that the lack of resolution of the water-stretching bands prevents their individual identity. The spectrum of calcium carbonate (calcite) is given as reference in Figure 8-23d. The sharp bands at 876 and 714 cm^{-1} are due to the carbonate as well as the broad peak at 1421 cm^{-1}.

The spectra for the zinc-doped samples are shown in Figure 8-24. Comparison of these reveals much about the effects of zinc addition. In Figure 8-24a the spectrum for the air-cured sample shows that the system reacts with carbon dioxide to form both calcium and zinc hydroxide. The bands at 1421, 1352, 876, and 714 cm^{-1} clearly show the presence of calcium and zinc carbonate. The sharp peak at 1391 cm^{-1} is due to the nitrate. Comparing the silicate bands

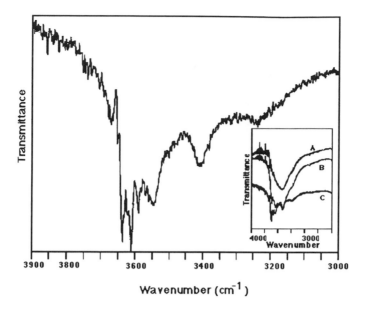

Figure 8-25.   Hydroxyl- and hydrogen-bonded bands in zinc-doped cement. (A) sample heated at 150°C for 2 h, (B) sample cured in air, and (C) computer-aided substraction of A from B.

in the region 930 to 995 cm^{-1} between Figure 8-23b and 8-24a show that the addition of zinc has retarded the formation of the silicate polymer. The sulfate band is clearly visible in the spectrum of the air cured zinc doped sample and it is not in the air-cured cement-only sample. Thus, zinc has prevented silicate polymerization. Figure 8-24b shows the zinc-doped sample cured in argon. The silicate region 930 to 995 cm^{-1} shows that the absence of carbon dioxide does not effect the spectrum compared to the air-cured sample as would be expected if the zinc prevents the polymerization of the silicate. The resolution of the water-stretching bands into six peaks is clearly visible and indicates that the water is prevented from forming the hydrated silicate products. The bands are due to calcium hydroxide, zinc hydroxide, surface bound hydroxides, and hydrogen-bonding hydroxides. The spectrum of $CaZn_2(OH)_6 \cdot 2H_2O$ is given in Figure 8-24d. The bands at 3652 and 3637 cm^{-1} are due to calcium and zinc hydroxides, respectively. Its similarity to the spectrum in Figure 8-24a of the air-cured Zn-doped cement sample supports the presence of the $CaZn_2(OH)_6 \cdot 2H_2O$.

The water-stretching region potentially holds the information needed to identify the nature of the active surface species, the hydroxyl species, the site specificity for adsorbing species that interact with hydroxyl groups, and the nature of the hydroxylated surface compound formed by a metal such as zinc. The problem has been that the region is composed of a large number of

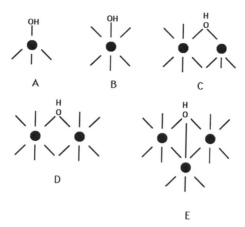

Figure 8-26.   Structures of surface hydroxides. Solid circles represent metals.

overlapping bands broadened by hydrogen bonding. FTIR spectra taken on zinc-doped samples cured in air and the same sample heated at 150°C for 2 h allowed a subtraction that produced a well-resolved water-stretch region with several broad bands between 3000 and 3580 cm^{-1} and five sharp bands between 3580 and 3700 cm^{-1} (Figure 8-25). The latter probably result from discrete surface hydroxyl species that are bound to the surface zinc and calcium compound. A similar situation has been noted on the surface of γ-alumina.[111-113] These structures are shown in Figure 8-26.

The basic character decreases from A to E. Consequently the most basic should have the highest vibrational frequency. The bands at this time cannot be assigned but future experiments may help identify the true nature of the bands. Nevertheless, the presence of a number of noninteracting OH groups has been identified and can now be used to study the nature of the surface species and their surface reactivity. The broad bands at lower wave numbers are most likely due to interacting surface and structural hydroxyls that have varying degrees of hydrogen bonding.

The chemistry of cement is complex and is made more so by the addition of metals. Therefore, we will concentrate our discussion on three main aspects of the problem. First, the hydration of the cement and the influence of the metal dopant, zinc, on this process is of primary importance. Here the nature of the water is critical. Second, the carbonation of the Portland cement and the influence of the zinc thereon is a developing new area of research with critical and broad implications for S/S systems. Finally, the state of polymerization of the silicate and influence of the zinc thereon is intimately related to the two issues above. In addition, the use of the OH bands in the 3800 to 3200 cm^{-1} to determine cation site occupancy is being explored. The results will then be discussed in terms of the binding sites and nature of the incorporated metals.

It has been proposed that zinc retards the hydration and setting of cement by the surface precipitation of an amorphous layer of zinc hydroxide on the anhydrous clinker grains and that calcium hydroxide reacts with the zinc hydroxide to form $CaZn_2(OH)_6 \cdot 2H_2O$.[109,110] Our FTIR data strongly support this proposal. However the mechanism by which this reaction takes place is not clear. The hydrating cement clinker should rapidly become negatively charged in the high pH being generated in the contacting solution. Cations will adsorb but anions such as $Zn(OH)_3^{-1}$ and $Zn(OH)_4^{2-}$ will be rejected. Thus, the participation of these ions can be ruled out unless their product with the reaction of Ca^{2+} such as $CaZn(OH)_3^+$ and $CaZn(OH)_4$ can be invoked. There is probably plenty of Ca^{2+} at the surface to provide for such a pathway. Since the pH is in the range that such Zn species will be present, as indicated by Figure 8-6, it seems quite likely that this type of positively charged calcium-containing species may be involved. The two waters of hydration indicate that water can be physically involved in the amorphous surface coating. A similar amorphous water containing Fe-oxide material[114] that resists material transport has been found to be important in the passivation layer on iron. This amorphous Zn compound can function in a similar fashion preventing material (water and ion) transport necessary for the hydration of the cement clinker.

In previous work[18] we have noted that adding zinc produces enhanced carbonate formation. The formation of the $CaZn_2(OH)_6 \cdot 2H_2O$ can act to prevent the formation of the tobermorite structure that has the $Ca(OH)_2$ layer of octahedra lying between two silicate sheets. Thus, the $Ca(OH)_2$ that forms will be more accessible to carbon dioxide for carbonate formation. In addition the surface zinc hydroxide can act to form carbonate as well. The carbonate effect needs to be delineated.

The polymerization of silicate can be followed and used in the study of cement-based polymerization processes. The observed polymerization of silicate on reaction of hydrating cement with carbon dioxide shows that the C-S-H-CO_2 reaction can compete with the CH-CO_2 reaction. The reaction of carbon dioxide with C-S-H thus can significantly account for the silicate polymerization observed. However, the mechanism has not yet been established. The release of water by the carbonation reaction can enhance the cement hydration reaction and subsequent silicate polymerization. Since leaching with acid enhances silicate polymerization[8,17] the process needs to be well understood to improve the S/S of hazardous metals.

The present results confirm our previous reports[11] that lead is preferentially adsorbed on the surface of the cementitious material doped with lead and also retardation of the setting of cement by Pb^{2+} ions by blocking the hydration reaction. The binding energy of the Pb 4f XPS peaks indicates the presence of Pb^{2+} ions and is also consistent with silicate, hydroxide, and carbonate. The leaching process removes most of the Pb^{2+} ions from the surface of the cement matrix as is evidenced from the much-reduced intensity of the Pb 4f XPS peak. The situation is similar for the zinc-doped sample. The presence of flakey materials on the surface of unleached samples and the low intensity of the lead and zinc observed in the EDS support surface binding for both lead and zinc. The SEM micrographs also support this argument.

The binding energy of the $Ca(2p_{3/2})$ peaks for the unleached and leached samples of both lead and zinc are not significantly different and could be attributed to Ca^{2+} in $Ca(OH)_2$. The surface concentration of calcium in the leached samples has been significantly reduced. Since calcium and silicon are the major components in Portland cement, a rough calculation for comparison purposes of the relative percentage of calcium and silicon in the leached and unleached samples can be made from the EDS spectra (Figure 8-22) after normalization of calcium and silicon peak areas only. In the unleached samples the percent of silicon and calcium are 29.0 and 71.0 for 10% Pb-doped cement and 19.1 and 80.9 for 10% Zn-doped cement, respectively. The corresponding values in the leached samples are 49.4 and 50.6 for Pb-doped cement and 83.7 and 16.3 for 10% Zn-doped cement. The Ca/Si ratio (2.40 and 1.02 for unleached and leached Pb-doped cement; 4.2 and 0.19 in the unleached and leached Zn-doped cement) thus decreases due to leaching, although the effect is more significant for Zn-doped cement. This is clear evidence that much of the calcium present as calcium hydroxide has been dissolved into the acidic medium. The relative concentration of silicon has been increased due to leaching and loss of hydroxide and sulfate.

The chemical shift with respect to the Si 2p XPS peak is a significant result. McWhinney et al.[100] have reported the binding energy, BE, of Si 2p XPS peaks in standard cement, 10% Pb-doped and 10% Zn-doped cement as 101.1, 101.6, and 101.5 eV, respectively. The BE of the Si 2p XPS peak in the unleached lead- and zinc-doped cement are comparable, while the BE of the same peak in the leached sample was found to be 102.7 and 102.6 eV, respectively. The BE of the Si 2p peaks have, therefore, been shifted in the leached samples by about 1.5 eV for lead and by 1.6 eV for zinc. The results clearly demonstrate that the silicon environment has undergone drastic changes due to leaching. We have reported[54-56] via XPS studies that the chemical shift with respect to Si 2p peak can be attributed to the difference in the Si-bonding environment and the BE of the resulting Si-containing compound is dependent on the number of Si-O bonds per Si atom. Calcium silicate hydrate ($CaO·SiO_2·H_2O$) is the primary cementing and hardening chemical formed on hydration in Portland cement and other lime silica systems. It is believed[11] that CaO is a favorable site for acid attack during leaching process leading to polymerization of the silicates. Ortego et al.[102] have reported via ^{29}Si NMR and FTIR studies of lead- and zinc-doped cement that polymerization of the silicates present in the cement matrix occurs due to leaching and the following generalized mechanism of polymerization has been proposed:

$$> Si - OX + H^+_{(aq)} \rightarrow > Si - OH + X^+_{(aq)} \text{ or}$$

$$> Si - O_2X + 2H^+_{(aq)} \rightarrow > Si - (OH)_2 + X^{2+}_{(aq)} \text{ etc.}$$

where X = calcium, potassium, sodium, or toxic metals ions. Then, condensations via silanol groups produces

$$Si - (OH)_x \rightarrow \text{branched and cross-linked silicates}$$

A BE value of 103.0 eV for Si 2p XPS peaks in the leached samples clearly demonstrates that the Si moiety in the cement matrix has been polymerized due to leaching. It is likely that small units of polymerized silica have been linked at one or two corners of O-Si-O tetrahedra. The beehive-type morphology present in the leached front of the SEM micrographs supports this argument. The presence of trace amounts of potassium in the unleached and leached samples is attributed to the ion exchange mechanism previously reported by us.[11]

Cadmium-Portland Cement (Surface Deposition)

Cadmium, toxic[115-119] to most living organisms, has received little attention with respect to S/S. Bishop[120] has investigated the leaching behaviors of synthetic metal hydroxide sludges containing Cd (0.2 M), Cr (0.46 M), and Pd (0.11 M) solidified by Portland cement (OPC) and discussed the possible mechanisms of metal binding. Other authors[121-123] have reported the leaching characterstics of cadmium and cadmium-containing sludges doped in OPC and fly ash-blended OPC, under different leaching conditions. More recent work by Roy et al.[124] have used SEM, EDX, optical spectroscopy, and XRD to evaluate the mechanisms of solidification/stabilization (S/S) of heavy metal sludges containing Cd, Cr, Hg, and Ni by OPC/fly ash binders. Our results show that Cd is located mainly on the outer surface of the C-S-H and its location and chemical nature is responsible for its retarding of cement hydration.

Speciation of Cd from FTIR

Our efforts[8-19] on the characterization of Cd-S/S systems have yielded some direct information on Cd speciation. Typical FTIR spectra of the Type V dry clinkers (OPC), hydrated OPC, and Cd-OPC are presented in Figure 8-27. A close examination of the silicate bands between Figures 8-27a and 8-27b reveal much of the effect of hydration on the polymerization of the silicate groups in OPC. The decrease in the intensity of the Si-O asymmetric stretching band due to tricalcium silicate (C_3S) and β-dicalcium silicate (β-C_2S) centered at around 900 cm^{-1} and the growth of a broad and strong band centered at 970 cm^{-1} has been attributed due to the formation of CaO·SiO$_2$·H$_2$O (C-S-H). The Si-O band at 925 cm^{-1} in the dry clinkers exists in the hydrated system and the new band due to hydrated silica phases appear at much higher frequency (centered at 960

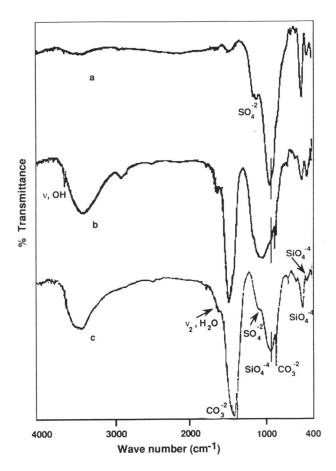

Figure 8-27.   FTIR spectra of Portland cement type V: (a) dry clinkers, (b) hydrated OPC cured in air, and (c) OPC-Cd cured in air.

cm^{-1}) in the air-cured sample. The silicate band has, thus, been shifted by 35 wave number units due to polymerization.

The sulfate bands at 1160 to 1100 cm^{-1} (Figure 8-27a) have now been obscured due to polymerization of the orthosilicate units. Comparison of the relative intensities of v_4 and v_2 silicate bands (Figure 8-27a and b) also indicate the degree of polymerization. Note the decrease in intensity of v_4 SiO_4^{4-}, (out-of-plane Si-O bending) vibration appearing at 525 cm^{-1}, and the relative increase of the v_2, SiO_4^{4-} (in-plane Si-O bending) vibration centered at 458 cm^{-1}. The sharp band in the water-stretching region at 3650 cm^{-1} is due to the OH stretch of the $Ca(OH)_2$. The poorly resolved strong and broad peak (Figure 8-27b) centered around 1430 cm^{-1} and the sharp peak at 875 cm^{-1} are due to carbonates.

However, when OPC is hydrated in the presence of metal dopant cadmium the v_3 SiO_4^{4-} band now appears at 935 cm^{-1} and the relative intensities of the v_4 SiO_4^{4-} and v_2 SiO_4^{4-} bands (Figure 8-27c) have also been reversed as compared with those in the OPC air-cured sample (Figure 8-27b) only. The sulfate bands have also reappeared. It, thus, indicates lower degree of polymerization of the orthosilicate units due to retardation of the hydration of OPC by cadmium. Note the virtual absence of OH bands in the water region and the increase in intensities of the carbonate bands at 1430, 876, and 713 cm^{-1}. The prominence of the carbonate bands and the absence of OH bands indicate that calcium hydroxide (CH) reacts preferentially with atmospheric CO_2 to C-S-H according to the reaction, $Ca(OH)_2 + CO_2 \rightarrow CaCO_3 + H_2O$, as proposed by a number of authors.[81,125] The sharp band appearing at 1386 cm^{-1} is due to nitrate ions from the cadmium salt used.

The FTIR spectra of OPC and Cd-OPC cured in an argon-rich atmosphere indicate that the silicate region is centered at around 930 cm^{-1} and that the absence of carbon dioxide does not affect the spectrum compared to the air-cured sample as would be expected if the cadmium prevents the polymerization of the silicate. The resolution of the water-stretching bands into a number of peaks in the argon-cured sample was clearly visible and indicates that the water is prevented from forming the hydrated silicate products. These peaks are not visible in the air cured Cd-OPC sample only. However, the Si-O band at 925 cm^{-1} in the dry clinkers no longer exists and the new band due to silica phases appears at much higher frequency field centered at 960 cm^{-1} in the air-cured sample. The silicate band has, thus, been shifted by 35 wave number units due to polymerization. This is possibly due to relatively higher rates of reactions of atmospheric carbon dioxide with the the hydroxides as mentioned in the previous section. However, the bands are due to calcium hydroxide, cadmium hydroxide, surface bound hydroxides, and hydrogen-bonding hydroxides. The bands at 3650 and 3635 cm^{-1} are due to calcium and cadmium hydroxides, respectively. A computer-simulated subtraction of the FTIR spectra taken on Cd-doped sample cured in argon-rich atmosphere and the same sample heated at 200°C for 2 h produced a well-resolved water-stretch region with several bands in the region 3650 to 3000 cm^{-1} (Figure 8-28). The sharp peaks at 3620 to 3525 cm^{-1} are attributed to surface-bound hydroxides and the bands appearing between 3490 to 3225 cm^{-1} are considered to be due to hydrogen-bonding hydroxides. The surface hydroxides probably result from discrete surface hydroxyl species that are bound to the surface cadmium and calcium compound. A similar situation has been noted on the surface of γ-alumina.[106-108] In a recent publication[15] we have reported the presence of a number of surface hydroxyl bands in zinc-doped OPC due to the formation of amorphous $CaZn_2(OH)_6 \cdot 2H_2O$ on the surface of cement grains.

The FTIR spectra of the leached Cd sample is shown in Figure 8-29b. The FTIR spectrum of the corresponding unleached sample (Figure 8-29a) is also included for comparison purpose. A broad and strong band due to hydrated

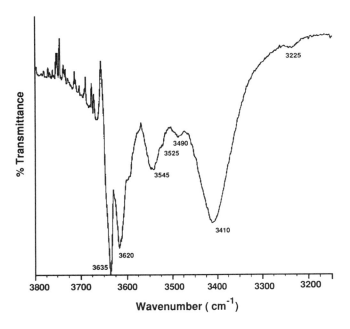

Figure 8-28.   **Hydroxyl- and hydrogen-bonded bands in cadmium-doped Portland cement.**

silica phases appears at around 1050 cm^{-1} with a shoulder at 1190 cm^{-1}. The shifting up of the v_3 SiO$_4^{4-}$ band by 100 cm^{-1} and the changes of the relative intensities of the v_4 SiO$_4^{4-}$ and v_2 SiO$_4^{4-}$ bands is indicative of a higher degree of polymerization of the orthosilicate units facilitated by the dissolution of the cadmium-containing complexes due to exposure to the acidic leaching solution.

The FTIR silicate and water regions provide information about the mutual interactions of the dopant metals and the cement particles and have broad implications on the S/S systems. We shall, therefore, concentrate our discussion on these two aspects. The solidification/stabilization of cadmium will be greatly influenced by the aqueous chemistry of Cd^{2+} ions. It has been reported that over $4 < $ pH $ < 8$, the predominant species in solution are Cd^{2+} ions and, above pH $= 8$, Cd(OH)$_2$ may be precipitated by alkalies. However, in a highly alkaline environment, like cement paste (pH $= 11$ to 13), hydroxocadmiates such as Cd(OH)$_4^{2-}$ will be formed. The adsorption reactions of Cd will depend on the kinds and amounts of Lewis acids and bases, (oxides and hydroxides in cement paste), the redox status (pe) and the acidity alkalinity (pH) of the particular environments.[92,93]

FTIR results indicate that the presence of cadmium has retarded the hydration of OPC. If one closely examines the FTIR spectra of OPC and Cd-OPC it can be clearly seen that the relative intensities of the two bending bands

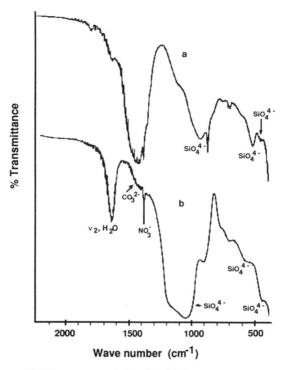

Figure 8-29. FTIR spectra of (a) Cd-OPC air cured and (b) Cd-OPC leached.

centered between 450 and 525 cm^{-1} in OPC have not been changed very much. The degree of hydration in OPC will affect the intensities of the bending modes ν_2 and ν_4 and it is generally expected that the intensities of these vibrations will decrease with increased polymerization. The major band, ν_3 SiO_4^{4-}, has been shifted by only about 30 wave number units in the Cd-OPC sample as compared with 76 cm^{-1} units in OPC sample, thus indicating the lower degree of polymerization.

The presence of several bands for surface hydroxyls near the water-stretching region is due to diverse coordination of the metal atoms bonded to the hydroxyl oxygen and is compatible with the formation of hydroxycadmiate type of complexes. The hydroxyl oxygen atoms may be bonded to one, two, and three lattice atoms. The formation of coordination bonds with metal atoms results in a lowering of stretching vibrational frequency and coordination with increasing number of metal atoms and is expected to lower the frequencies of vibrations. Previous reports[9,100] have claimed that the retardation of cement hydration by cadmium is due to the precipitation of $Cd(OH)_2$ on the surface of cement grains. We believe that any $Cd(OH)_2$ is transformed into cadmiates according to the following reaction:

$$Cd(OH)_2 + OH^- \rightarrow [Cd(OH)_4]^{2-}$$

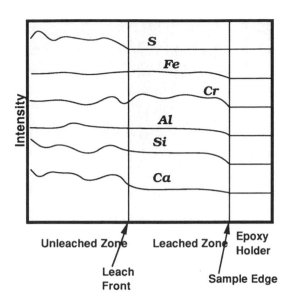

**Figure 8-30.   EDS line scans of leached 10% chromium-doped Port-
land cement.**

This species reacts with surface calcium to produce Ca[Cd(OH)$_4$]. It is probably this species on the surface that retards the cement setting.

The v_3 SiO$_4^{4-}$ band in the leached sample (Figure 8-29) appears as an
intense and broad peak centered at around 1060 cm^{-1} and the relative intensities of v_2 and v_4 have been reversed due to leaching. The v_2 peak is now
centered at 460 cm^{-1}, while the v_4 is barely visible as a shoulder. These
vibrational changes are indicative of increased polymerization of the orthosilicate
units in Cd-OPC air-cured sample due to the dissolution of cadmium complexes from the surface of the cement matrix when exposed to the acidic
leaching solution.

8.8. LEACH TESTING

Optimization of the S/S system from both physical and chemical immobilization is the design goal. The mechanism of leaching is needed to optimize
the S/S systems. SEM/EDS line scans across the leach front as shown in Figure
8-30 can provide much insight into the mechanism. The sharp front indicates
that the leaching is very dependent on the the loss of alkalinity in the cement.

The Ca leaches from the hydroxide but remains in the silicates. The sulfur
leaches almost completely from the system. The Al and Si profiles are smoothed
due to the redistribution of the silicate and aluminate structures. Cr and Fe
appear to increase in concentration but this is an artifact due to density loss. The

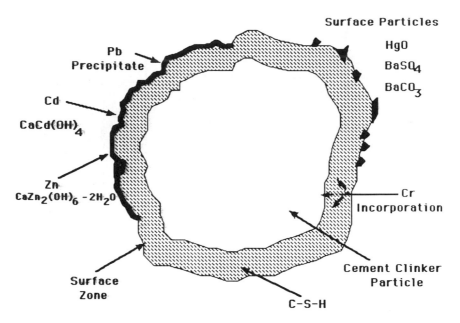

Figure 8-31.   A summary of models for the interaction of priority metal pollutants with cement.

detailed profiles at the leach front are important to model the leaching mechanisms. Future work should seek to produce higher resolution profiles in this region. Clearly, the calcium hydroxide is dissolved and the silicate structures are changed toward higher polymerization as shown by the chemical characterization results already discussed.

The surface acid-base chemistry coupled to the solution chemistry is a major consideration in leaching. Modeling the process requires that both the thermodynamic equilibria be assessed and the kinetic mechanisms be delineated. The latter needs the nature of the surface complex. This is as yet unavailable. Study of the relevant surfaces in actual aqueous systems is essential. New techniques using FTIR, AFM, and others will be required to detail the surface structures. Standard techniques for measuring leaching provides only an observed diffusivity that does not separate the chemical and physical factors. The observed diffusivity that can be calculated from modeling has been shown to depend on the true effective diffusivity and the coefficients that describe the chemistry involved.[126] The modeling of the leaching which should separate the influences of the physical and chemical aspects will ultimately provide reliable projections and design parameters that will ensure long term protection of the environment. Leaching in both cases causes substantial changes in the cement pore structure indicative of the actual dissolution of the cement as is suggested by the loss in strength caused by leaching.

8.9. CONCLUSIONS

Considerable progress has been made in the characterization of the binding states, the chemical and physical mechanisms of interaction and the leaching mechanisms of priority metal pollutants with cement. The results are formulated into models and summarized in Figure 8-31. XPS, ISS, and SEM/EDS results confirm that zinc, cadmium, and lead are preferentially deposited on the surface of the cement grains. FTIR has provided insight into the chemical nature of the surface species in the Cd and Zn system. Here the surface compounds have been identified as mixed hydroxides $CaCd(OH)_4$ and $CaZn_2(OH)_6 \cdot 2H_2O$, respectively. These species apparently result from the Ca-aided adsorption of the the normally anionic Cd and Zn species at high pH. The Pb surfaces species have yet to be identified but they are likely to be negative species in solution at high pH and will require careful study to determine the nature of the adsorbing entities. Hg has been proven to be present as surface particulate, HgO. The Ba has been found to be present as the sulfate and the carbonate. Cr is incorporated into the C-S-H matrix in some yet to be determined fashion.

One of the main findings of this study is the resolution of five sharp IR bands between 3580 and 3000 cm^{-1}. These bands have been attributed to the surface hydroxide species bound to the surface zinc and calcium and as well are due to hydrogen bonding in the water region. The FTIR spectra support the presence of $CaZn_2(OH)_6 \cdot 2H_2O$. However, further work in identifying the individual peaks and the nature of the bonding of different cations with the components of cement matrix is needed and is in progress. The results of this study have established that the retardation of the hydration of OPC is due to the formation of $CaCd(OH)_4$ on the surface of cement grains.

The results have established that the FTIR technique combined with XPS can be used as an effective combination of tools in monitoring the progress of complex chemical reactions during hydration of cement including to study the effect of dopant metal cations as well as carbonation on the hydration process. In particular the polymerization of silicate can be followed. The FTIR results also indicate that the atmospheric carbon dioxide and as well as the dopant metal greatly influence the polymerization of the silicates.

The leaching experiments using XPS, FTIR, and SEM/EDS reveal that the initial attack of the leaching solution is on the CH which is followed by attack on the C-S-H. Subsequent to C-S-H loss of calcium, silicate polymerization toward silica occurs. This produces a solid with much-reduced mechanical strength. The decrease in alkalinity caused by the acidic attack increases the solubility of the hazardous cations as well as undermines the chemical and physical integrity of the cement matrix.

The ideas put forward in this work and the experimental results point toward substantial future progress. The application of the best surface and bulk characterization techniques to the chemical and physical aspects of stabilization and solidification will be required if the complex systems are to be mastered to the extent needed for general design improvements.

ACKNOWLEDGMENTS

We wish to thank the Gulf Coast Hazardous Substance Research Center, Lamar University — Beaumont for primary support. Partial support for work related to the amorphous materials involved comes from the Welch Foundation, Houston. Special thanks go to Ricardo Davis, Hylton Mcwhinney, and Yung-Nien Tsai for their experimental work and to Tom Hess and Don Mencer for surface analysis and helpful discussions.

REFERENCES

1. Pojasek, R. B. 1979. "Solid-Waste Disposal: Solidification." *Chem. Eng.* 86(17): 141–145
2. Tittlebaum, M. W., R. K. Seals, F. K. Cartledge, and F. Engels. 1986. *Crit. Rev. Environ. Control* 15: 179.
3. Landreth, R. E. 1980. "Guide to the Disposal of Chemically Stabilized and Solidified Wastes." U.S. Environmental Protection Agency, Cincinnati (EPA SW-872).
4. Pojasek, R. B., Ed. 1987. *Toxic and Hazardous Waste Disposal*, Vol. 1 and 2. Ann Arbor Science Publishers, Ann Arbor, MI.
5. Conner, J. R. 1990. *Chemical Fixation and Solidification of Hazardous Wastes.* Van Nostrand Reinhold, New York.
6. Barth, E. F., Ed. 1990. *Stabilization and Solidification of Hazardous Wastes.* Noyes Data Corporation, Park Ridge, NJ.
7. Cote, P. and Gilliam, M., Eds. 1989. *Environmental Aspects of Stabilization and Solidification of Hazardous and Radioactive Wastes.* American Society for Testing and Materials, Philadelphia.
8. Cocke, D. L. 1990. "The Binding Chemistry and Leaching Mechanisms of Hazardous Substances in Cementitious Solidification/Stabilization Systems." *J. Haz. Materials* 24: 231–253.
9. Ortego, J. D., S. Jackson, G.-S. Yu, H. G. McWhinney, and D. L. Cocke. 1989. "Solidification of Hazardous Substances — A TGA and FTIR Study of Portland Cement Containing Metal Nitrates." *J. Environ. Sci. Health*, 24: 589–602.
10. Cocke, D. L., J. D. Ortego, H. G. McWhinney, K. Lee, and S. Shukla. 1989. "Model for Lead Retardation of Cement Setting." *Cement Concrete Res.* 19: 156–159.

11.   Cocke, D. L., H. G. McWhinney, D. C. Dufner, B. Horrell, and J. D. Ortego.1989. "An XPS and EDS Investigation of Portland Cement Doped with Pb(II) and Cr(III) Cations." *Haz. Waste Haz. Materials* 6: 251–267.

12.   McWhinney, H. G., D. L. Cocke, L. K. Balke, and J. D. Ortego. 1990. "An Investigation of Mercury Solidification in Portland Cement Using X-Ray Photoelectron Spectroscopy and Energy Dispersive Spectroscopy." *Cement Concrete Res.* 20(1): 79.

13.   McWhinney, H. G., M. W. Rowe, D. L. Cocke, J. D. Ortego, and G. S. Yu. 1990. "X-Ray Photoelectron and FTIR Spectroscopic Investigation of Cement Doped with Barium Nitrate." *Environ. Sci. Health* A25(5): 463–477.

14.   Cocke, D. L., M. Y. A. Mollah, J. R. Parga, and G. Z. Wei. 1991. "Binding Chemistry and Leaching Mechanisms in Solidified Hazardous Wastes." Proc. 3rd Annu. Symp. Gulf Coast Hazardous Substance Research Center, Beaumont, TX, Feb. 21–22, pp.196–197.

15.   Davis, R. C. and D. L. Cocke. 1991. "Analysis of Physical and Chemical Aspects of Leaching Bevior in Lead- and Chromium-Doped Portland Cement." Proc. 5th Int. Symp. on Ceramics in Nuclear and Hazardous Waste Management, Cincinnati, April 28–May 2, pp. 1–9.

16.   McWhinney, H. G., D. L. Cocke, and L. Donaghe. 1990. "Surface Characterization of Priority Metal Pollutants in Portland Cement." Proc. Hazardous Materials Control Resources Incorporated's 7th National Resource Conservation and Recovery Act/Superfund Conference on Hazardous Waste and Hazardous Materials, St. Louis, May 2–4.

17.   Cocke, D. L. 1990. "The Binding Chemistry and Leaching Mechanisms of Hazardous Substances in Cementitious Solidification/Stabilization Systems." Proc. Gulf Coast Hazardous Substance Research Center Conference on Solidification /Stabilization, Mechanisms and Applications.

18.   Mollah, M. Y. A., J. R. Parga, and D. L. Cocke. in press. "An Infrared Spectroscopic Examination of Cement Based Solidification/ Stabilization Systems — Portland Type V and Type IP with Zinc." *J. Environ. Sci. Health.*

19.   Cocke, D. L., M. Y. A. Mollah, J. R. Parga, T. R. Hess, and J. D. Ortego. 1992. "An XPS and SEM/EDS Characterization of Leaching Effects on Pb and Zn Doped Portland Cement." *J. Haz. Materials.* 30: 83–95.

20.   Gribble, C. D. and A. J. Hall. 1985. *Practical Introduction to Mineralogy.* Atten and Unwin, Wichester, MA.

21.   Watt, I. M. 1985. *The Principles and Practice of Electron Microscopy.* Cambridge University Press, New York.

22.   Diamond, S., J. F. Young, and F. V. Lawrence, Jr. 1986. "Scanning Electron Microscopy-Energy Dispersive X-Ray Analysis of Cement Constituents — Some Cautions." *Cement Concrete Res.* 4: 899; Joy, D. C., A. D. Romig, and J. I. Goldstein, Eds. *Principles of Analytical Electron Microscopy.* Plenum Press, New York.

23.   Brownmiller, L. T. and R. H. Bogue. 1930. *Am. J. Sci.* 20: 241; Culity, B. D. 1956. *Elements of X-Ray Diffraction.* Addison-Wesley, Reading, MA.

24.   Jenkins, R., R. W. Gould, and D. Gedcke. 1981. *Quantitative X-Ray Spectrometry.* Marcel Dekker, New York.

25.   Mirabella, F. M., Jr. 1985. "Internal Reflection Spectroscopy." *Appl. Spectrosc. Rev.* 21(1): 45–178.

26.   Barrett, E. P., L. G. Joyner, and P. C. Halenda. 1951. "Determination of Pore Volume and Area Distribution in Porous Substances. I. Computation from N Isotherm." *J. Am. Chem. Soc.* 73: 373.

27.   Griffiths, P. R. 1975. *Chemical Infrared Fourier Transform Spectroscopy.* John Wiley & Sons, New York.

28.   Bensted, J. and S. P. Varma. 1974. "Infrared and Raman Spectroscopy of Cement Chemistry." *Cem. Tech.* 5: 378.

29.   Bensted, J. 1976. "Examination of the Hydration of Slag and Pozzolanic Cement by Infrared Spectroscopy." *Cemento* 73(4): 209–214.

30.   Regourd, M., J. H. Thomassin, P. Baillif, and J. C. Touray. 1980. "Study of the Early Hydration of Ca_3SiO_5 by X-Ray Photoelectron Spectrometry." *Cem. Concr. Res.* 10: 223.

31.   Nakamota, K. 1978. Infrared and Raman Spectra of Inorganic and Coordination Compounds, J. Wiley & Sons, New York.

32.   Rouxhet, P. G. 1970. "Hydroxyl Stretching Bands in Micas: A Quantitative Interpretation." *Clay Minerals* 8: 375–388.

33.   Sanz, J. and W. E. E. Stone, W. E. E. 1983. "NMR Study of Minerals. III. The Distribution of Mg^{2+} and Fe^{2+} Around the OH Groups in Micas." *J. Phys.* C16: 1271–1281.

34.   Aines, R. D. and G. R. Rossman. 1984a. "Water in Minerals? A Peak in the Infrared." *J. Geophys. Res.* 89: 4059–4071.

35.   Aines, R. D. and G. R. Rossman. 1984b. "The High Temperature Behavior of Water and Carbon Dioxide in Cordierite and Beryl." *Am. Mineral* 69: 319–327.

36.   Aines, R. D. and G. R. Rossman. 1984c. "Water Content of Mantle Garnets." *Geology* 12: 720–723.

37.   Aines, R. D. and G. R. Rossman. 1984d. "The Hydrous Component in Garnets: Pyralspites." *Am. Mineral* 89: 1116–1126.

38.   Aines, R. D. and G. R. Rossman. 1985. "The High Temperature Behavior of Trace Hydrous Components in Silicate Minerals." *Am. Mineral* 70: 1169–1179.

39.   Farmer, V. C. 1974. "Layer Silicates" in: *Infrared Spectra of Minerals.* Farmer, V. C., Ed. Mineral Society, London.

40. Gilkes, R. J., R. C. Young, and J. P. Quirk. 1972. "The Oxidation of Octahedral Iron in Biotite." *Clays Clay Minerals* 20: 303–315.
41. Herbillon, A. J. and J. Tran Vinh. 1969. "An, Amorphous Colloidal Fraction of Some Tropical Soils." *J. Soil Sci.* 20: 223.
42. Vempati, R. K., R. H. Leoppert, and D. L. Cocke. 1990. "Mineralogy and Reactivity of Amorphous Si-Ferrihydrites, Solid State Ionics." 38: 53–61.
43. Kirkpatrick, R.J. 1988. "MAS-NMR Spectroscopy of Minerals and Glasses in Spectroscopic Methods in Mineralogy and Geology." Hawthorne, F. C., Ed. Mineralogical Socierty of America, 341–403.
44. Mering, H. 1982. *Principles of High Resolution NMR in Solids.* Springer-Verlag, New York.
45. Barnes, J. R., A. D. H. Clague, N. J. Clayden, C. M. Dobson, C. J. Hayes, G. W. Groves, and S. A. Rodger. 1984. "Hydration of Portland Cement Followed by ^{29}Si Solid-State NMR Spectroscopy." *J. Mater. Sci. Lett.* 4: 1293; *Chem. Commun.* 1396.
46. Lippmaa, E., M. Magi, M. Tarmak, W. Wieker, and A. R. Grimmer. 1982. "A High Resolution 29Si NMR Study of the Hydration of Tricalciumsilicates." *Cement Concrete Res.* 12: 597.
47. Clayden, N. J., C. N. Dobson, C. A. Hayes, and S. A. Rodger. 1984. "Hydration of Tricalciumsilicates Followed by Solid-State ^{29}Si NMR Spectroscopy." *J. Chem. Soc. Chem. Commun.* 1396.
48. Lippmaa, E., A. A. Madis, T. J. Penk, and G. Engelhardt. 1978. "Solid-State High Resolution NMR Spectroscopy of Spin 1/2 Nuclei (^{13}C, ^{29}Si, ^{119}Sn) in Organic Compounds." *J. Am. Chem. Soc.* 100: 1929.
49. Cartledge, F. K., L. G. Butler, D. Chalasani, H. C. Eaton, F. P. Frey, E. Herrera, M. E. Tittlebaum, and S.-L. Yag. 1990. "Immobilization Mechanisms in Solidification/Stabilization of Cadmium and Lead Salts Using Portland Cement Fixing Agents." *Environ. Sci. Technol.* 24:867.
50. Briggs, D. and M. P. Seah, Eds. *Practical Surface Analysis by Auger and Photoelectron Spectroscopy.* John Wiley & Sons, New York.
51. Seah, N. P. and Briggs, D. 1983. *Practical Surface Analysis by Auger and Photoelectron Spectroscopy.* Briggs, D. and M. P. Seah, Eds. John Wiley & Sons, New York.
52. Horrell, B. A. and D. L. Cocke. 1987. "Applications of Ion-Scattering Spectroscopy to Catalyst Characterization." *Catal. Rev. Sci. Eng.* 29(4): 447–491.
53. Chuah, G. K. and D. L. Cocke. 1986. "The Study of Metal Oxidation by Surface Sensitive Mass Spectroscopic Techniques." *J. Trace Microprobe Techniques* 4(1 and2): 1–36.
54. Vempati, R. K., R. H. Loeppert, D. C. Dufner, and D. L. Cocke. 1990. "X-ray Photoelectron Spectroscopy as a Tool to Differentiate Silicon-Bonding State in Amorphous Iron Oxides." *Soil Sci. Soc. Am. J.* 54(3): 695–698.

55.   Cocke, D. L., R. K. Vempati, and R. H. Loeppert. " Applications of XPS to the Study of Soil Chemistry and Soil Mineralogy." ASA publication, Madison, WI. Accepted for publication.

56.   Vempati, R. K., D. L. Cocke, and J. W. Stucki. "X-Ray Photoelectron Spectroscopy of Clay and Clay Minerals." D. Reidel, Dordrecht, The Netherlands. To be published.

57.   Koppleman, M. H. 1979. *Advanced Chemical Methods for Soil and Clay Minerals.* Stucki, J. W. and W. L. Banwart, Eds. D. Reidel, Dordrecht, The Netherlands.

58.   Carriere, B., J. P. Deville, D. Brion, and J. Escard. 1977. "X-ray Photoelectron Study of Some Silicon-Oxygen Compounds." *J. Electron Spectrosc. Relat. Phenom.* 10: 85.

59.   Lea, F. M. 1971. *The Chemistry of Cement and Concrete,* 3rd Ed. Chemical Publishing, New York; Bogue, R. H. 1955. *The Chemistry of Portland Cement,* 2nd Ed. Van Nostrand Reinhold, New York.

60.   Laitinen, H. A. and W. E. Harris. 1987. *Chemical Analysis.* McGraw Hill, New York.

61.   West, A. R. 1987. *Solid State Chemistry and Its Implications.* John Wiley & Sons, New York.

62.   Cotton, F. A., G. Wilkinson, and P. L. Gaus. 1987. *Basic Inorganic Chemistry.* John Wiley & Sons, New York. p. 620.

63.   (a)   Thomas, N. L., D. A. Jameson, and D. D. Double. 1981. "The Effect of Lead Nitrate on the Early Hydration of Portland Cement." Cement Concrete Res. 11: 143.
       (b)   Double, D. D., N. L. Thomas, and D. A. Jameson. 1980. "The Hydration of Portland Cement. Evidence for an Osmotic Mechanism." Proc. 7th Int. Congress on the Chemistry of Cement.

64.   Locher, F. W. 1979. *Zement Taschenbuck.* Bauverlag, Weisbaden. p. 19.

65.   Hansen, J. 1982. "The Delicate Architecture of Cement." *Science* 49.

66.   Tashiro, C. 1980. Proc. 7th Cong. Chem. Cement, Vol. 2. Paris. pp.11–37.

67.   Tashiro, C., H. Takahashi, M. Kanaya, I. Hirakida, and R. Yoshida. 1977. "Hardening Property of Cement Mortar Adding Heavy Metal Compounds and Solubility of Heavy Metals from Hardened Mortar." *Cement Concrete Res.* 7: 283.

68.   Double, D. D. 1978. *Silicates Industriels* 11: 233.

69.   Birchall, J. D., A. J. Howard, and J. E. Bailey. 1978. "On the Hydration of Portland Cement." *Proc. Roy. Soc. London* A360 : 445.

70.   Birchall, J. D., A. J. Howard, and D. D. Double. 1980. "Some General Considerations of a Membrane/Osmosis Model for Portland Cement Hydration." *Cement Concrete Res.* 10: 145 (1980).

71.   Beckett, D. 1983. "Influence of Carbonation and Chlorides on Concrete Durability." *Concrete* 17(2): 16–18.

72. Cole, W. F. and B. Kroone. 1959. "Carbonate Minerals in Hydrated Portland Cement." *Nature* No. 4688.

73. Dayal, R. and R. Klein. 1972. "Evaluation of Cement Based Backfill Materials." Ontario Hydro Research Division, Report 87-223-K.

74. Sauman, Z. 1972. "Effect of CO_2 on Porous Concrete." *Cement Concr. Res.* 2: 541–549.

75. Kondo, R., D. Masaki, and T. Akiba. 1968. "Mechanisms and Kinetics on Carbonation of Hardened Cement." Proc. 5th Int. Symp. Chem. Cement, Tokyo, 402–409.

76. Suzuki, K., T. Nishikawa, and S. Ito. 1985. "Formation and Carbonation of CSH in Water." *Cement Concr. Res.* 15, 213–224.

77. Baird, T. G., A. G. Cairns-Smith, and D. S. Snell. 1975. "Morphology and CO_2 Uptake in Tobermorite Gel." *J. Colloid Interface Sci.* 50:387–391.

78. Calleja, J. 1980. "Durability". Proc. 7th Int. Symp. Chem. Ceme., Paris, 1–43.

79. Smolczyk, H. G. 1968. "Written Discussion — Carbonation." Proc. 5th Int. Symp. Chem. Cement, Toyko, 369–384.

80. Kamimura, K., P. J. Sereda, and E. G. Swenson. 1965. "Changes in Weight and Dimensions in the Drying and Carbonation of Portland Cement Mortars." *Magaz. Conc. Res.* 17: 5–14.

81. Venuat, M. and J. Alexandre. 1968. "De la Carbonation du Beton." *Rev. Mater. Const.* 421–481.

82. Young, J. F., R. L. Berger, and J. Breese. 1974. "Accelerated Curing of Compacted Calcium Silicate Mortars on Exposure to CO_2." *J. Am. Ceram. Soc.* 57: 394–397.

83. Fukushima, T. 1984. "Chemical Behavior of Calcium Hydroxide in Cement Paste and Concrete." Proc. 27th Jap. Cong. Mater. Res., 225–232.

84. Steinour, H. H. 1958. "The Setting of Portland Cement: A Review of Theory, Performance and Control." *Portland Cement Assoc. Res. Dev. Bull.* 98: 124.

85. Lieber, W. 1967. "Effects of Zinc Oxide on the Setting and Hardening of Portland Cement." *Zem.-Kalk-Gips* 56(3): 91–95.

86. Edwards, G. C. and R. L. Angstadt. 1966. "Effects of Some Soluble Inorganic Admixtures on the Early Hydration of Portland Cement." *J. Appl. Chem.* 16:166.

87. Mindess, S. and J. F. Young. 1981. *Concrete*. Prentice Hall, Englewood Cliffs, NJ.

88. Powers, T. C. 1962. "A Hypothesis on Carbonation Shrinkage." *J. Port. Cement Assoc.,* 4: 40–50.

89. Slegers, P.A. and P. G. Rouxhet. 1976. "Carbonation of the Hydration Products of Tricalcium Silicate." *Cement Concrete Res.* 6: 381–388.

90. Stumm, W., H. Hohl, and F. Dalang. 1976. "Interactions of Metal Ions with Hydrous Oxide Surfaces." *Croat. Chem. Acta* 48 (4): 491–504.

91. Stumm, W., C. P. Huang, and S. R. Jenkins. 1970. "Specific Chemical Interaction Affecting the Stability of Dispersed Systems." *Croat. Chem. Acta* 42: 223–245.

92. Stumm, W. and J. J. Morgan. 1981. *Aquatic Chemistry,* 2nd Ed. John Wiley & Sons, New York.

93. Schindler, P. W. and W. Stumm. 1987. *Aquatic Surface Chemistry.* Stumm, W., Ed. John Wiley & Sons, New York. p. 83.

94. Schindler, P. W. 1981. *Adsorption of Inorganics at Solid-Liquid Interfaces.* Anderson, M. A. and Rubin, A., Eds. Ann Arbor Science, Ann Arbor, MI. p. 1.

95. Schindler, P. W., B. Fuerst, R. Dick, and P. U. Wolf. 1976. "Ligand Properties of Surface Silanol Groups. I. Surface Complex Formation With Fe^{3+}, Cu^{2+}, Cd^{2+} and Pb^{2+}." *J Colloid Interface Sci.* 55: 469.

96. Benjamin, M. M. and J. O. Leckie. 1981. "Adsorption of Metals at Oxide Surfaces: Effects of the Concentrations of Adsorbate and Competing Metals." *Environ. Sci. Technol.* 15: 1050.

97. Corey, R. B. 1981. *Adsorption of Inorganics at Solid-Liquid Interfaces.* Anderson, M. A. and Rubin, A., Eds. Ann Arbor Science, Ann Arbor, MI.

98. Hingston, F. J. 1981. *Adsorption of Inorganics at Solid-Liquid Interfaces.* Anderson, M. A. and Rubin, A., Eds. Ann Arbor Science, Ann Arbor, MI. p. 51.

99. Greenwood, N. N. and A. Earnshaw. 1984. *Chemistry of the Elements.* Pergamon Press, Elmsford, NY. pp. 1190–1199.

100. McWhinney, H. G. 1990. "Surface Characterization of Priority Metal Pollutants in Portland Cement." Ph.D. dissertation, Texas A & M University.

101. Davis, R. C. 1990. Masters thesis, Texas A & M University.

102. Ortego, J. D., Y. Barroeta, F. K. Cartledge, and H. Akhter. 1991. Leaching Effects on Silica Polymerization. An FTIR and ^{29}Si NMR Study of Lead and Zinc in Portland Cement." *Environ. Sci. Technol.* 25:1171–1174.

103. (a) Komarneni, S., E. Breval, D. M. Roy, and R. Roy. 1988. "Reactions of Some Calcium Silicates with Metal Cations, *Cement Concrete Res.* 18: 204–220.

    (b) Miyake, M., S. Komarneni, and R. Roy. 1988. "Immobilization of Pb^{2+}, Cd^{2+}, Sr^{2+} and Ba^{2+} Ions Using Calcite and Aragonite." *Cement Concrete Res.* 18, 485–490.

104. Saguma, T. 1987. "The Effect of $Ca(OH)_2$ Treated Ceramic Microspheres on the Mechanical Properties of High Temperature Lightweight Cement Composites." *Material Research Soc. Symp. Proc.* 114: 301–307.

105. Tashiro, C. and J. Oba. 1979. "The Effects of Chromium(III) Oxide, Copper(II) Hydroxide, Zinc Oxide and Lead(II) Oxide on The Compressive Strength and Hydrates of Hardened C3A Paste." *Cement Concrete Res.* 9(2), 253–258.

106. Tashiro, C., J. Oba, and K. Akama. 1979. "The Effects of Several Heavy Metal Oxides on The Formation of Ettringite and Microstructure of Hardened Ettringite." Cement Concrete Res. 9: 303–308.

107. Takahashi, H., M. Shinkado, I. Hirakida, and S. Hasegawa. 1973. *Semento Gijutsu Nenpo.* 17: 91.

108. Takahashi, H., M. Shinkado, I. Hirakida, and K. Uto. 1974. *Semento Gijutsu Nenpo.* 28: 121.

109. Arliguie, G., J. P. Ollivier, and J. Grandet. 1982. "Etude de L'effet Retardateur du Zinc Sur L'Hydration de la Pate de Ciment Portland." *Cement Concrete Res.* 12: 79–86.

110. Arliguie, G. and J. Grandet. 1990. "Influence de la Composition D'Un Ciment Portland Sur Son Hydratation en Presence de Zinc." *Cement Concrete Res.* 20: 517–524.

111. Peri, J. B. and R. B. Hannan. 1960. "Surface Hydroxyl Groups on γ-Alumina." *J. Phys. Chem.* 69: 220.

112. Peri, J. B. 1965. "A Model for the Surface of γ-Alumina." *J. Phys. Chem.* 69(1): 220.

113. Oblad, A. G., T. H. Milliken, and G. A. Mills. 1951. "Chemical Characteristics and Structures of Cracking Catalysis." *Adv. Catal.* 3:199–247.

114. Murphy, O. J., J. O'M. Bockris, T. E. Pou, D. L. Cocke, and G. Sparrow. 1982. "SIMS Evidence Concerning Water in Passive Layers." *J. Electrochem. Soc.* 129: 2149–2145.

115. Webb, M. 1979. *The Chemistry, Biochemistry and Biology of Cadmium.* Webb. M., Ed. Elsevier/North-Holland, Amsterdam. pp 285–340.

116. Bermner, I. 1974. *The Chemistry, Biochemistry and Biology of Cadmium.* Webb., M., Ed. Elsevier/North-Holland, Amsterdam. pp 285–340.

117. Jastrow, D. J. and E. D. Koeppe. 1980. *Cadmium in the Environment, Part 1,* Nriagu, J. O., Ed. John Wiley & Sons, New York. pp 607–655.

118. Flinders, C. G. 1982. "Cadmium and Health: A Survey." *Int. J. Environ. Stud.* 19(3–4): 187–193.

119. Bretherick, L., Ed. 1981. *Hazards in Chemical Laboratory,* 3rd Ed. Royal Society of Chemistry, London.

120. Bishop, P. C. 1988. "Leaching of Inorganic Hazardous Constituents from Stabilized/Solidified Hazardous Wastes." *Haz. Waste Haz. Materials* 2(5): 1229.

121. Melkinger-Campbll, K., T. El-Korchi, D. Gress, and P. C. Bishop. 1987. "Stabilization of Cadmium and Lead in Portland Cement Paste Using a Synthetic Seawater." *Environ. Progress* 2(6): 99.

122. Poon, C. S. and R. Perry. 1987. In *Materials Research Society Symposia Proceedings: Fly Ash and Coal Conversion By-Products Characterization, Utilization, and Disposal II.* G. J. McCarth, et al., Eds., Materials Research Society, Pittsburgh. p. 67.

123.   Shin, H.-S., J.-K. Koo, J.-O. Kim, and S.-P. Yoon. 1990. "Leaching
       Characteristics of Heavy Metrals from Solidified Sludge under Seawa-
       ter Conditions." *Haz. Waste Haz. Materials* 7(3): 261–271.
124.   Roy, A., H. C. Eaton, F. K. Cartledge, and M. E. Tittlebaum. 1991.
       "Solidification/Stabilization of a Heavy Metal Sludge by a Portland
       Cement/Fly Ash Binding Mixture." *Haz. Waste Haz. Materials* 8(1):
       33–41.
125.   Reardon, E. J., B. R. James, and J. Abouchar. 1989. "High Pressure
       Carbonation of Cementitious Grout." *Cement Concrete Res.*19: 385–
       399.
126.   Batchlor, B. 1989. "Modeling Chemical and Physical Processes in
       Leaching Solidified Wastes." 3rd Int. Conf. New Frontiers for Hazard-
       ous Waste Management, U.S. Environmental Protection Agency, Cin-
       cinnati (EPA/600/989-072).

9 EFFECTS OF EQUILIBRIUM CHEMISTRY ON LEACHING OF CONTAMINANTS FROM STABILIZED/SOLIDIFIED WASTES

B. Batchelor and K. Wu

9.1. ABSTRACT

Simple leaching models demonstrate the importance of both chemical and physical factors on leaching of contaminants from solidified/stabilized (s/s) wastes. Chemical mechanisms reduce mobility by reducing the fraction of contaminant that is in a mobile phase. In many cases, these chemical immobilization reactions can be assumed to be in metastable equilibrium. A chemical equilibrium model called SOLTEQ is being developed which can predict the equilibrium speciation in wastes treated by s/s. It is based on the chemical equilibrium program MINTEQA2 and has important components from the programs SOLMINEQ and SIMUL. It is capable of calculating activity coefficients at the high ionic strength found in pore water of wastes treated by s/s. It has formation coefficients for many species identified in pozzolanic systems, and it is capable of describing the variable stoichiometry of calcium silicate hydrate (C-S-H), which is main product of pozzolanic reactions. The model provides a flexible framework for describing equilibrium chemistry in wastes treated by s/s and can be linked to leaching models to improve their ability to describe movement of contaminants under a variety of conditions.

0-87371-748-1/93/$0.00+.50
© 1993 by Lewis Publishers

9.2. INTRODUCTION

s/s is an increasingly important technology for treatment of hazardous, radioactive, and mixed wastes. The first priority in managing these wastes should always be to reduce the amounts produced and then to recycle as much as possible. In most cases, application of the best waste reduction and recycle technologies will still result in some waste that needs to be disposed. s/s technologies can then be used to treat the waste so that the release of contaminants to the environment is minimized.

Both physical and chemical mechanisms work in solidified wastes to immobilize contaminants. The physical mechanism is associated with entrapping the contaminants within a solid matrix, thereby reducing their ability to move in the environment. The extent of immobilization by physical means depends on the characteristics of the pore structure in the treated waste. As such, it is a characteristic of the waste form, and is the same for all contaminants. Because physical immobilization depends on the microstructure of the solid matrix, it is affected by the chemical reactions that produce that matrix.

Chemical immobilization mechanisms are associated with those chemical reactions that change a contaminant from a more to a less mobile form. Precipitation, sorption, and oxidation/reduction are important types of reactions that result in chemical immobilization.

The relative importance of the kinetic and equilibrium characteristics of these reactions to the degree of immobilization depends upon the conditions within the waste form. The kinetics of immobilization reactions could be important if their rates are about the same speed or slower than the rate of physical transport of the contaminant through the matrix. In most cases, this will not be true because most chemical immobilization reactions are controlled by the rate of diffusion of a mobile contaminant to or from the matrix surface. Immobilization processes will be much faster than the physical process of diffusion through the solid matrix because the path for diffusion through the matrix is the pore length, which will almost always be longer than the path for diffusion that leads to chemical reaction, which is the pore radius. Therefore, kinetics will generally not play a major role in chemical immobilization mechanisms.

Equilibrium characteristics of immobilization reactions will generally determine the extent of chemical immobilization. However, it is important to distinguish between true equilibrium conditions and metastable equilibrium in these systems. True equilibrium would be reached when all reactions reach their global free energy minimum and the net rate of change becomes zero. Metastable equilibrium is reached when the net rate of reaction approaches zero for a significant time, but is not necessarily equal to zero. This is the condition expected in many solidified wastes after an initial curing period results in stable conditions. However, this is not true equilibrium, because slow changes in the composition of the solid matrix and the pore water constituents can continue for years.

The purpose of this paper is to review how chemical and physical immobilization mechanisms can be incorporated into leaching models and how knowledge of the equilibrium chemistry of binding agents and contaminants can be used to develop a model of their equilibrium chemistry that can be applied to describing chemical immobilization in solidified wastes.

9.3. LEACHING MODELS

A variety of different mechanistic leaching models can be derived from a material balance on a contaminant in a waste form. They will differ based on the assumptions made about the geometry of the waste form, chemical reactions of the contaminant, and conditions under which leaching occurs. The role of equilibrium chemistry in defining leaching behavior will be demonstrated here by assuming the simplest conditions. For a semi-infinite solid of rectangular geometry contained within a well-mixed, infinite bath, the material balance equation and its initial and boundary conditions are as follows.

Initial condition: $C = C_{t0}$ at $t = 0$, all X

$$\frac{\partial C}{\partial t} = D_e \frac{\partial^2 C}{\partial X^2} - R \tag{1}$$

Boundary conditions: $C = 0$, at $X = 0$, all t

$$C \rightarrow C_{t0} \text{ as } X \rightarrow \infty$$

where C = concentration of mobile component
   C_{t0} = concentration of mobile component at time zero
   t = time
   D_e = effective diffusivity
   X = distance into solid
   R = rate of removal of mobile component

If the contaminant does not react, this system can be solved to give the following relationship for the fraction of contaminant leached:[1]

$$\frac{M_t}{M_o} = \left(\frac{4D_e}{\pi L^2}\right)^{0.5} t^{0.5} \tag{2}$$

where M_t = mass of component that has leached at time = t
   M_0 = mass of component in solid at time = 0
   L = distance from center of slab to its surface

The effect of simple reactions on leaching can be considered easily if the reactions are assumed to be in equilibrium. This will normally be a good assumption, since the time required for reaction should generally be small compared to the time required for transport out of the waste form. For a number of simple reactions, the leaching equation can be shown to be of the same form as Equation 2, but containing an observed diffusivity rather than the effective diffusivity.[2,3] The definition of the observed diffusivity depends on the reactions that are assumed to occur.

$$\frac{M_t}{M_o} = \left(\frac{4D_{obs}}{\pi L^2} \right)^{0.5} t^{0.5} \tag{3}$$

If linear sorption is assumed to occur, the observed diffusivity is defined by the following.

$$D_{obs} = \frac{D_e}{1 + K_p} = F_m D_e \tag{4}$$

where D_{obs} = observed diffusivity
K_p = linear partition coefficient, equal to ratio of concentration of sorbed phase to concentration of solution phase at equilibrium
F_m = fraction of contaminant that is initially mobile

If a portion of the contaminant is assumed to have reversibly precipitated, the following definition for observed diffusivity applies:

$$D_{obs} = \frac{\pi [F_m (1 - F_m) + 0.5 F_m^2] D_e}{2} \tag{5}$$

If the fraction of contaminant in the mobile (dissolved) phase is small, this reduces to the following:

$$D_{obs} = \frac{\pi F_m D_e}{2} \tag{6}$$

For the cases presented here, the observed diffusivity is nearly equal to the effective diffusivity times the fraction of contaminant that is mobile. This rule of thumb also has been found to be approximately valid during numerical simulations of nonlinear sorption.[4] These results demonstrate the importance of reducing the fraction of contaminant in the mobile phase in order to reduce leaching, as others have emphasized.[5] These simple models also demonstrate the importance of equilibrium chemistry to leaching because it determines the distribution of a contaminant between mobile and immobile forms which determines the fraction of contaminant that is mobile.

The simple reactions assumed for the leach models described above cannot describe all of the many complex chemical interactions that can occur in wastes treated by s/s. Many components in the binder and the waste interact chemically. For example, binder hydration products maintain a high concentration of hydroxide in pore water and this results in precipitation of many contaminants as hydroxides. Contaminants could also react with the solid matrix and each other. For example, lead could sorb onto the calcium silicate hydrate surface in a similar manner as it sorbs onto the surface of other oxides.[6] Lead could also react with chromate ion to form a precipitate. Components from the leaching solution can also react with binder and waste components. For example, products of binder hydration provide substantial acid-neutralizing capacity to resist attack from acids in the leaching media. This potentially complex reaction system cannot be generally described by simple models that can be solved analytically. Numerical solution techniques are required to solve these complex, multicomponent models. Such a multicomponent leaching model would also need a general model to describe equilibrium partitioning of the binder and waste components. A general equilibrium model should contain a database of thermodynamic data that is applicable to compounds found in waste forms and should be able to apply them to conditions typically found in the pore water of wastes treated by s/s. Such a model could also serve as a framework to coordinate knowledge on the chemistry of contaminants and binders, even when it is not used to predict leaching.

9.4. BINDER CHEMISTRY

A number of binders have been used to solidify and stabilize waste materials, but portland cement, fly ash, kiln dust, blast furnace slag, silica, and lime are probably the most common. The chemistry of portland cement and its hydration products has been studied extensively due to the use of portland cement as a construction material. The primary hydration products of portland cement are a calcium silicate hydrate (C-S-H), lime, tetracalcium aluminate $\{2[Ca_2Al(OH)_7 \cdot 3H_2O]\}$, ettringite $\{Ca_6[Al(OH)_6]_2(SO_4)_3 \cdot 26H_2O\}$, monosulfate $\{Ca_4[Al(OH)_6]_2SO_4 \cdot 6H_2O\}$, hydrogarnet $\{Ca_3[Al(OH)_6]_2\}$, and thaumassite $\{Ca_6[Si(OH)_6]_2(CO_3)_2(SO_4)_2 \cdot 24H_2O\}$.[7] The presence of most of these has been confirmed by X-ray diffraction.[8]

The kinetics of hydration of Portland cement has been described as occurring in five stages.[9] The first four stages are complete within a day or two. The last stage is a slow reaction that can continue for many years. Analysis of pore water in hydrating cement pastes shows that the concentration of alkali metals increases for about 30 d, then remains constant for months.[10] Other pore water constituents including pH have been observed to have the same behavior.[11]

Therefore, the complex chemical system of hydrating Portland cement does not reach true equilibrium for years, but a metastable equilibrium is often reached after about a month of hydration.

Some investigators have applied tools of equilibrium chemistry to study the metastable phase of cement hydration. The importance of ionic strength of pore water in describing the precipitation equilibrium of calcium hydroxide (portlandite) has been reported.[10,12,13] This equilibrium could be described by a solubility product when activity corrections were made by the Debye-Hückel method and ionic strength was below 0.2 M. Correcting the solubility product with another method improved its ability to predict pore water composition.[13] The most important solid phase in hydrated cement is a calcium silicate hydrate (C-S-H) gel, which is known to have variable stoichiometry. The equilibrium behavior of this material has been summarized as falling into two classes, one for C-S-H formed soon after hydration begins and the other for C-S-H formed later.[14,15] This approach has been used to develop empirical equations to describe the stoichiometry and solubility of C-S-H.[16] The Pitzer ion interaction model[17] was used in this development to describe the effects of ionic strength on the equilibrium chemistry. Information on the solid species present in solidified wastes is another requirement of general chemical equilibrium models. One model has been developed that is based on assuming the primary solids to be C-S-H, lime, monosulfate, gehlenite hydrate, and hydrotalctite.[8]

A variety of other pozzolanic materials are used as binders with Portland cement or by themselves. Fly ash and silica fume are the binders that have been studied most extensively, due to their use in construction materials. Hydration products of these binders have been reported to be similar to those for Portland cement.[18,19] However, it is difficult to generalize about the behavior of binders such as fly ash because of its chemical and physical variability.[20]

Fly ash is known to react more slowly that Portland cement.[18,20,21] Unreacted fly ash particles have been found after 30 to 90 d of curing.[20] Slower reaction is assumed due to layers of hydration products that hinder diffusion.[8,22] Therefore, metastable equilibrium may not occur as rapidly when binders such as fly ash are used to solidify wastes.

Addition of fly ash and silica fume to Portland cement causes some changes in the composition of hydration products. Less lime is produced and the Ca/Si ratio in C-S-H is reduced.[8,20,22] The content of aluminum, potassium, and sodium in C-S-H increases in systems with fly ash.[20] However, fly ash does not significantly change many of the pore water concentrations. A slight increase in sodium and a slight decrease in potassium have been observed when fly ash is added to Portland cement pastes.[23] Others report a decrease in hydroxide and chloride in pore water when fly ash, silica fume, or blast furnace slag are added to Portland cement.[24]

Chemical behavior of contaminants in solidified wastes has received much less study, but it is increasing. Lead has been identified as primarily existing as a mixed precipitate of hydroxide, sulfate, and other components in a solidified matrix.[25] Mercury is believed to exist primarily in a soluble form at lower concentrations[26] or partially as an oxide precipitate when concentrations are high.[27] Cadmium is believed to primarily exist as a hydroxide precipitate.[25,26] Zinc also has been reported to be precipitated at high pH.[26] Barium is believed to be converted to sulfate and carbonate precipitates in solidified matrices.[28] Trivalent chrome maintains its oxidation state in solidified wastes and has characteristics that are consistent with its presence as silicate, carbonate, or hydroxide precipitate.[29] Strontium is reported to replace calcium in C-S-H, calcium sulfate precipitate, and ettringite.[30] Technetium normally is found as the soluble pertechnate ion.[31] However, when sufficient blast furnace slag is added to reduce technetium to the technetium dioxide form, observed diffusivities can be reduced by as much as five orders of magnitude.[31]

9.5. DEVELOPMENT OF A MODEL FOR EQUILIBRIUM CHEMISTRY

Models for leaching of contaminants from solidified wastes demonstrate the need for a method to describe speciation of contaminants in the multicomponent environment of solidified wastes. Such a model should be able to describe equilibrium chemistry of binder components as well as waste components. It should be able to describe the effect of high ionic strengths on equilibrium chemistry and it should be able to incorporate solid phases known to exist in these systems. Furthermore, it should be flexible so that it could be applied to many different systems consisting of various binders and wastes.

In addition to providing a method to predict speciation of contaminants in solidified waste forms, a chemical equilibrium model would also provide a framework for interpreting experimental results. Comparing model predictions to experimental data would provide a method of evaluating the validity of the assumptions that form the model. Assumptions could be modified and the model's ability to predict results could again be tested.

SOLTEQ is a model that is being developed to describe chemical equilibrium in solidified wastes. The chemical equilibrium model MINTEQA2 was used as the basis for this new model, but components of other models such as SOLMINEQ[32] and SIMUL[16] also have been incorporated. SOLTEQ was constructed from MINTEQA2 by changing its technique for calculating activity coefficients, by adding thermodynamic data for binder hydration products to its database, and by modifying its calculation procedure to account for the variable stoichiometry of C-S-H.

The Pitzer ion-interaction model[17] was chosen to calculate activity corrections for major binder components because of its ability to accurately predict activity coefficients at high ionic strengths. Sufficient data was available to apply the Pitzer method to the following species: Ca^{2+}, Na^+, K^+, Mg^{2+}, $MgOH^+$, H^+, Al^{3+}, $Al(OH)_4^-$, Cd^{2+}, OH^-, Cl^-, SO_4^{2+}, HSO_4^-, CO_3^{2-}, HCO_3^-, $H_3SiO_4^-$, $H_2SiO_4^{2-}$, CrO_4^{2-}, and $Fe(OH)_4^-$. Use of the ion interaction model is not compatible with calculating concentrations of weakly bound ion pairs.[16,32]

Therefore, the following species were eliminated from consideration by SOLTEQ: $MgCO_3^0$, $MgHCO_3^+$, $MgSO_4^0$, $CaHCO_3^+$, $CaCO_3^0$, $CaSO_4^0$, $NaCO_3^-$, $NaHCO_3$, $NaSO_4^-$, KSO_4^-, $CdSO_4^0$, $Cd(SO_4)_2^{2-}$, $NaCrO_4^{2-}$, $KCrO_4^{2-}$, $NaAc^-$, $CaOH^+$, $Cd(SO_4)_2^{2-}$, $AlSO_4^+$, and $Al(SO_4)_2^-$. The B· method[33] was used to calculate activity coefficients for those components for which ion interaction coefficients were not available. The B· coefficient was calculated separately for ions with one, two, or three charges in such a way that the activity coefficients for Na^+, Ca^{2+}, and Al^{3+} would be the same whether calculated by the B· method or by the Pitzer method. This approach extends the range of the B· method to ionic strengths of 3 to 6 M.[32]

$$\log \gamma_i = \frac{-Z_i^2 A_\gamma \sqrt{I}}{1 + a_i^0 B_\gamma \sqrt{I}} + B_i \cdot I \qquad (7)$$

where

γ_i = activity coefficient for species i
Z_i = charge on species i
A_γ, B_γ = Debye-Hückel coefficients,
I = ionic strength
a_i^0 = ion size parameter
B· = deviation function

$$B_1 \cdot = \frac{1}{I}\left(\log \gamma_{Na^+} + \frac{A_\gamma \sqrt{I}}{1 + 4 \times 10^{-8} B_\gamma \sqrt{I}} \right) \qquad (8)$$

$$B_2 \cdot = \frac{1}{I}\left(\log \gamma_{Ca^{2+}} + \frac{4 A_\gamma \sqrt{I}}{1 + 6 \times 10^{-8} B_\gamma \sqrt{I}} \right) \qquad (9)$$

$$B_3 \cdot = \frac{1}{I}\left(\log \gamma_{Al^{3+}} + \frac{6 A_\gamma \sqrt{I}}{1 + 9 \times 10^{-8} B_\gamma \sqrt{I}} \right) \qquad (10)$$

where γ_{Na+}, γ_{Ca2+}, γ_{Al3+} = activity coefficients for Na^+, Ca^{2+}, and Al^{3+} calculated by Pitzer method.

The activities of uncharged aqueous species were computed using the sum of the molality of all species in solution.[32,34]

$$\log \gamma = 0.00489 \times 10^{\frac{280}{T}} \cdot \sum_i Z_i^2 m_i \qquad (11)$$

**Table 9-1. Comparison of Activity Coefficients Calculated by
SOLTEQ and SIMUL**

Component	Activity Coefficient		
	SOLTEQ	SIMUL	Difference (%)
Ca^{2+}	0.079	0.074	6.5
$MgOH^+$	0.596	0.598	0.3
Mg^{2+}	0.126	0.114	10.0
Na^+	0.716	0.717	0.1
K^+	0.767	0.767	0.0
H^+	0.603	0.605	0.3
OH^-	0.686	0.689	0.4
$Al(OH)_4^-$	0.653	0.655	0.3
$Al(OH)_3^0$	1.077	—	
$Al(OH)_2^+$	0.756	—	
$AlOH^{2+}$	0.074	0.072	2.7
Al^{3+}	0.009	0.004	76.9
$H_2SiO_4^{2+}$	0.165	0.136	19.3
$H_3SiO_4^-$	0.653	0.655	0.3
$H_4SiO_4^0$	1.077	1.045	3.0

where T = absolute temperature (°K)

m_i = molality of species i.

Table 9-1 shows good agreement between activity coefficients calculated by SOLTEQ and SIMUL. SIMUL also uses the ion-interaction method to calculate activities.[16]

The equilibrium chemistry of C-S-H was incorporated into SOLTEQ using an empirical approach[16] that is based on data compiled from many experiments on C-S-H.[14,15] The regression equation for the formation constant of C-S-H was modified from that presented for the solubility product[16] to make it compatible with MINTEQA2.

$$(Ca \mathbin{/} Si)_{csh} = 0.48548 + 0.11563*R + 0.0104536*R^2$$

$$-\log K_{csh} = -5.0165 - 1.9066*R - 0.2735*R^2$$

where R = logarithm of the ratio of the activity of Ca^{2+} to the activity of $H_4SiO_4^0$ in the solution

K_{csh} = formation constant for C-S-H

Table 9-2. Species Added to the MINTEQA2 Database

Name	Formula	Log K
Ettringite	$Ca_6Al_2O_6(SO_4)_3 \cdot 32H_2O$	−58.862
Ca-carboaluminate	$CaO \cdot Al_2O_3 \cdot CaCO_3 \cdot 11H_2O$	−26.100
Sepiolite	$Mg_4Si_6O_{15}(OH)_2 \cdot 6H_2O$	−32.625
C4AH13	$4CaO \cdot Al_2O_3 \cdot 13H_2O$	−102.510
C2AH8	$2CaO \cdot Al_2O_3 \cdot 8H_2O$	−60.956
C3AH16	$3CaO \cdot Al_2O_3 \cdot 16H_2O$	−82.042
Ca-Al-Si gel	$3CaO \cdot Al_2O_3 \cdot 2SiO_2$	−34.104
Ca-Oxychloride	$Ca_4Cl_2(OH)_6 \cdot 4H_2O$	−71.738
Mg-Oxychloride	$Mg_2Cl(OH)_3 \cdot 13H_2O$	−26.034
Syngenite	$K_2Ca(SO_4)_2 \cdot H_2O$	7.450
Monoaluminosulfate	$Ca_4Al_2O_6SO_4 \cdot 12H_2O$	−74.372
K-Al-Si gel	$K_2O \cdot Al_2O_3 \cdot 6SiO_2$	−19.442
C-S-H	$rCaO \cdot SiO_2 \cdot (1+r)H_2O$	−25.047
Alunogen	$Al(SO_4)_3 \cdot 17H_2O$	−6.200
Friedl Salt	$Ca_4Al_2Cl_2(OH)_{12} \cdot 4H_2O$	−86.012
Naheolite	$NaHCO_3$	10.730
Glaserite	$K_3Na(SO_4)_2$	3.800
Leonite	$K_2SO_4 \cdot MgSO_4 \cdot 4H_2O$	3.980
Schoenite	$K_2SO_4 \cdot MgSO_4 \cdot 6H_2O$	4.330
Polyhalite	$K_2SO_4 \cdot MgSO_4 \cdot 2CaSO_4 \cdot 2H_2O$	13.750
Pickeringite	$MgSO_4 \cdot Al_2(SO_4)_3 \cdot 22H_2O$	8.060
Halotrichite	$FeSO_4 \cdot Al_2(SO_4)_3 \cdot 22H_2O$	10.470
Arcanite	K_2SO_4	1.780

This model was incorporated into SOLTEQ and modifications were made to the algorithm used by MINTEQA2 to solve the set of material balance equations so that it could account for the variable stoichiometry of C-S-H.

Equilibrium constants[16] and stoichiometry for 22 solids often found in cementitious systems were added to the database for SOLTEQ. These are shown in Table 9-2.

Figure 9-1 shows how concentrations of calcium and silica computed by SOLTEQ agree with the trend line found for ten separate sets of experimental data.[14] The model is able to reproduce the trend line very well.

Figure 9-2 shows how SOLTEQ calculations compare to measured concentrations in pore water expressed from a Portland cement paste cured for 10 months.[35] If SOLTEQ provided perfect predictions, all points in this figure would lie on the horizontal line. Points above the line indicate that SOLTEQ predicted a concentration for that component that was higher than that measured. Points below the line indicate that SOLTEQ made a prediction that was

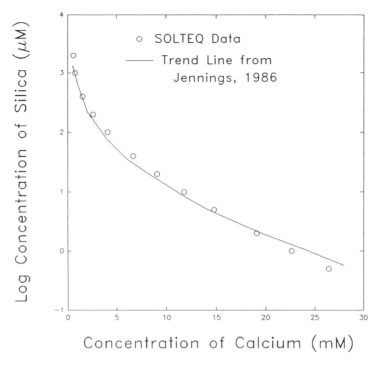

Concentration of Calcium (mM)

Figure 9-1.   **Comparison of silica and calcium concentrations calculated by SOLTEQ with trend of data summarized by
Jennings. (From Jennings, H. M., 1986. *J. Am. Ceram. Soc.*
69(8): 614–618. With permssion.)**

too low. At this stage of its development, SOLTEQ is unable to accurately
describe the partitioning of Na and K, so the data shown in Figure 9-2 were
obtained using measured concentrations of Na and K. Figure 9-2 shows generally good agreement between SOLTEQ and measured concentrations for
hydroxide (OH) and most of the major matrix components (Al, Ca, Mg, Si).
However, the predicted concentration for Fe is much too high. This may be due
to the lack of thermodynamic data for tetracalcium aluminoferrite in the model.

Figure 9-3 shows how SOLTEQ can predict concentrations of contaminants
in solidified wastes. The data for comparison to SOLTEQ was obtained from
a study that contacted ground-solidified waste with a small amount of water
until it was believed that equilibrium conditions were met.[5] SOLTEQ was able
to predict concentrations of hydroxide and major matrix components (Ca, Fe,
Si, Al) fairly well. SOLTEQ predicted concentrations of chromium (Cr) and
cadmium (Cd) reasonably well, but substantially underestimated the concentration of lead (Pb). This may be the result of SOLTEQ using a formation
coefficient for lead hydroxide that is representative of a crystalline solid, rather
than a coefficient that describes formation of an amorphous solid that might be

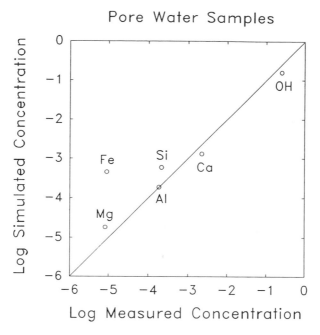

Figure 9-2.   Comparison of SOLTEQ predictions with measured concentrations in pore water from a Portland cement paste.

found during metastable equilibrium. Amorphous solids would be more soluble and would cause higher pore water concentrations. This demonstrates the need to obtain more information on the equilibrium chemistry of contaminants in pore waters of wastes treated by s/s.

SOLTEQ can be used to improve leaching models by calculating the concentration of contaminant that is present in a mobile form under conditions in the waste form. Contaminant that is present in an immobile form such as a precipitated solid cannot leach, but first must be converted to a mobile form in solution. Movement of the mobile contaminant through the solid can be modeled by diffusion theory. As the concentration of mobile contaminant is decreased within the solid by leaching, portions of the immobile phase can be converted to the mobile phase. Therefore, leach models incorporating chemical equilibrium must consider at a minimum two forms of the contaminant. These can be mobile and immobile or mobile and total (mobile plus immobile).

SOLTEQ is able to calculate the concentrations of a variety of species for each component when given the total concentrations of all components as input. This is a generic approach and can be used to describe behavior of any number of components in a waste treated by s/s. The concept of a component is taken fromMINTEQA2, where a component includes all species of an element at the same oxidation state. For example the component calcium can be present in a number of species such as the calcium ion (Ca^{2+}), calcium

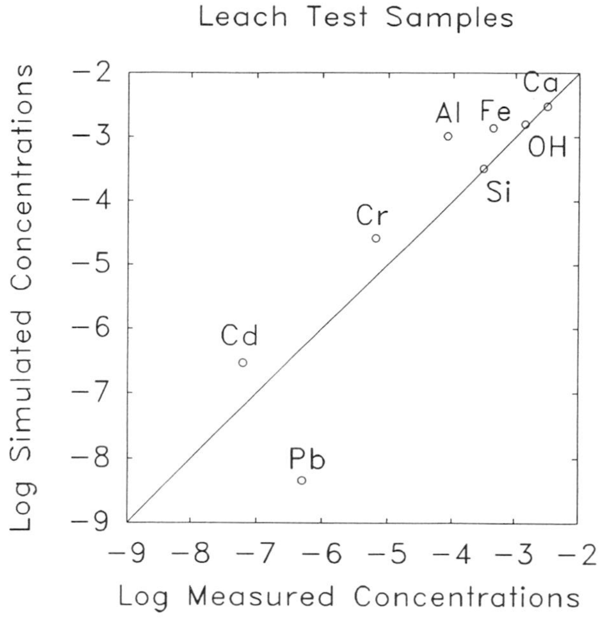

Figure 9-3.   **Comparison of SOLTEQ predictions with concentrations measured in equilibrium leach test.**

hydroxide-soluble complex ($CaOH^+$), portlandite crystals [$Ca(OH)_2$], and a variety of other forms.

One way to use SOLTEQ to predict leaching behavior is to estimate the fraction of contaminant that is in the mobile phase (F_m). As shown by Equations 5 and 6, the fraction mobile can be used to calculate the observed diffusivity for simple one-component systems. As a first approximation, the observed diffusivity could be taken to be the product of the fraction leached and the effective diffusivity. However, this approach neglects any chemical interactions among components of the waste form and these interactions are likely to be important in most leaching conditions. For example, the solubility of many metals will be determined by the pore water pH, which in turn will be determined in large part by the behavior of solids such as portlandite and calcium silicate hydrate.

SOLTEQ can also be used to predict the effect of multicomponent chemical interactions on leaching. The material balance equation presented for a nonreactive component (Equation 1) must be modified to consider a component that can exist in a number of different forms. The accumulation term must consider the total concentration and the transport term must consider only the mobile phase concentration. Since all reactions are conversions among species of the same component, the rate term is irrelevant. The material balance equation for any component in the waste form becomes the following.

$$\frac{\partial T_i}{\partial t} = D_e \frac{\partial^2 C_{m,i}}{\partial x^2} \tag{12}$$

where T_i = total concentration of component i

   $C_{m,i}$ = concentration of component i in all its mobile forms

This equation can be solved simultaneously for a number of components by numerical techniques. At each time step in the calculations, SOLTEQ would be used to calculate the concentration of all mobile species for each component. The sum of these concentrations would be $C_{m,i}$. SOLTEQ would require as input the total concentrations of each component at that point in the waste form (T_i).

SOLTEQ is a work in progress. Improved numerical procedures are being developed to speed convergence. Improving its accuracy in predicting speciation of contaminants depends on specifying more accurately the solid phases present after reaching metastable equilibrium, and developing more accurate estimates for their equilibrium constants. These coefficients may well not be the same as those for well-crystallized solids that are normally found in thermodynamic databases. Coefficients that better describe conditions during metastable equilibrium will be needed. Continued development of SOLTEQ offers the potential to provide a tool that is both predictive and interpretive.

ACKNOWLEDGMENT

This project has been funded in part with federal funds as part of the program of the Gulf Coast Hazardous Substance Research Center, which is supported under cooperative agreement R 815197 with the U.S. Environmental Protection Agency and in part with funds from the State of Texas as part of the program of the Texas Hazardous Waste Research Center and the Texas Advanced Technology Program. The contents do not necessarily reflect the view and policies of the U.S. EPA or the State of Texas nor does the mention of trade names or commercial products constitute endorsement or recommendation for use.

REFERENCES

1. Crank, J. 1975. *The Mathematics of Diffusion*. Clarendon Press, Oxford.
2. Batchelor, B. 1989. "Modeling Chemical and Physical Processes in Leaching Solidified Wastes." *3rd Int. Conf. on New Frontiers for Hazardous Waste Management*, U.S. Environmental Protection Agency (EPA/600/9-89-072).

3. Batchelor, B. 1990. "Leach Models: Theory and Application." *J Haz. Materials* 24(2 and3): 255–266.
4. Batchelor, B. 1989. "Orthogonal Collocation as a Solution Technique for Leaching Models that Incorporate Chemical and Physical Mechanisms for Immobilization." 197th National Meeting, American Chemical Society, Dallas.
5. Cote, P. 1986. "Contaminant Leaching from Cement-Based Waste Forms under Acidic Conditions." Ph.D. thesis, McMaster University, Hamilton, Ontario.
6. Hohl, H. and W. Stumm. 1976. "Interaction of Pb^{2+} with Hydrous γ-Al_2O_3." *J Colloid Interfacial Sci.* 55(2):281–288.
7. Bensted, J. 1983. "Hydration of Portland Cement" in *Advances in Cement Technology*. Ghosh, S. N., Ed. Pergamon Press, Oxford.
8. Atkins, M., D. E. Macphee, and F. P. Glasser. 1989. "Chemical Modeling in Blended Cement Systems." *Fly Ash, Silica Fume, Slag, and Natural Pozzolans in Concrete, Proc. 3rd Int. Conf.* American Concrete Institute, Detroit.
9. Jennings, H. M. 1983. "The Developing Microstructure in Portland Cement" in *Advances in Cement Technology*. Ghosh, S. N., Ed. Pergamon Press, Oxford.
10. Diamond, S. 1975. "Long-Term Status of Calcium Hydroxide Saturation of Pore Solutions in Hardened Cements." *Cement Concrete Res.*5:607–616.
11. Page, C. L. and O. Vennesland. 1983. "Pore Solution Composition and Chloride Binding Capacity of Silica-Fume Cement Pastes." *Materiaux Constructions* 16(91): 19–25.
12. Moragues, A., A. Macias, and C. Andrade. 1987. "Equilibria of the Chemical Composition of the Concrete Pore Solution. I. Comparative Study of Synthetic and Extracted Solutions." *Cement Concrete Res.* 17:173–182.
13. Moragues, A., A. Macias, C. Andrade, and J. Losada. 1988. "Equilibria of the Chemical Composition of the Concrete Pore Solution. II. Calculation of the Equilibrium Constants of the Synthetic Solutions." *Cement Concrete Res.* 18:342–350.
14. Jennings, H. M. 1986. "Aqueous Solubility Relationships for Two Types of Calcium Silicate Hydrate." *J. Am. Ceram. Soc.*69(8): 614–618.
15. Gartner, E. M. and H. M. Jennings. 1987. "Thermodynamics of Calcium Silicate Hydrates and their Solutions." *J. Am. Ceram. Soc.* 70(10): 743–749.
16. Reardon, E. J. 1990. "An Ion Interaction Model for the Determination of Chemical Equilibria in Cement/Water Systems." *Cement Concrete Res.* 20: 175–192.

17. Pitzer, K. S. 1979. "Theory: Ion Interaction Approach" in *Activity Coefficients in Electrolyte Solutions*. Pytkowicz, R. M., Ed. CRC Press, Boca Raton, FL.

18. Ramachandran, V. S. 1983. "Waste and Recycled Materials in Concrete Technology" in *Advances in Cement Technology*. Ghosh, S. N., Ed. Pergamon Press, Oxford.

19. Sersale, R. 1983. "Aspects of the Chemistry of Additions" in *Advances in Cement Technology*. Ghosh, S. N., Ed. Pergamon Press, Oxford.

20. Roy, D. M. 1989. "Fly Ash and Silica Fume Chemistry and Hydration." *Fly Ash, Silica Fume, Slag, and Natural Pozzolans in Concrete, Proc. 3rd Int. Conf.* American Concrete Institute, Detroit.

21. Fraay, A. L. A. and J. M. Bijen. 1989. "The Reaction of Fly Ash in Concrete." *Cement Concrete Res.*19:235–246.

22. Mehta, P. K. 1989. "Pozzolanic and Cementitious By-Products in Concrete — Another Look." *Fly Ash, Silica Fume, Slag, and Natural Pozzolans in Concrete, Proc. 3rd Int. Conf.* American Concrete Institute, Detroit.

23. Berry, E. E., R. T. Hemmings, W. S. Langley, and G. G. Carette. 1989. "Beneficiated Fly Ash: Hydration, Microstructure, and Strength Development in Portland Cement Systems." *Fly Ash, Silica Fume, Slag, and Natural Pozzolans in Concrete, Proc. 3rd Int. Conf.* American Concrete Institute, Detroit.

24. Byfors, K., C. M. Hansson, and J. Tritthart. 1986. "Pore Solution Expression as a Method to Determine the Influence of Mineral Additives on Chloride Binding." *Cement Concrete Res.* 16(5): 760–770.

25. Cartledge, F. K., L. G. Butler, D. Chalasani, H. C. Eaton, F. P. Frey, E. Herrera, M. E. Tittlebaum, and S. Yang. 1990. "Immobilization Mechanisms in Solidification/Stabilization of Cd and Pb Salts Using Portland Cement Fixing Agents." *Environ. Sci. Technol.* 24(6): 867–873.

26. Poon, C. S., C. J. Peters, and R. Perry. 1985. "Mechanisms of Metal Stabilization by Cement Based Fixation Processes." *Science Total Environ.* 41: 55–71.

27. McWhinney, H. G., D. L. Cocke, K. Balke, and J. D. Ortego. 1990. "An Investigation of Mercury Solidification and Stabilization in Portland Cement Using X-Ray Photoelectron Spectroscopy and Energy Dispersive Spectroscopy." *CementConcrete Res.* 20(1): 79–91.

28. McWhinney, H. G., M. W. Rowe, D. L. Cocke, J. D. Ortego, and G. Yu. 1990. "X-Ray Photoelectron and FTIR Spectroscopic Investigation of Cement Doped with Barium Nitrate." *J Environ. Science Health* A25(5): 463–477.

29. Cocke, D. L., H. G. McWhinney, D. C. Dufner, B. Horrell, and J. D. Ortego. 1989. "An XDS and EDS Investigation of Portland Cement Doped with Pb^{2+} and Cr^{3+} Cations." *Haz. Wastes Haz. Materials* 6(3): 251–267.

30.   Atkinson, A., A. K. Nickerson, and T. M. Valentine. 1983. "The Mechanism of Leaching from Some Cement-Based Nuclear Wasteforms." U.K. Atomic Energy Authority, Harwell (AERE-R-10809) from P. Columbo, R. Doty, D. Dougherty, M. Fuhrman, and Y. Sanborn. 1985. "Leaching Mechanisms of Solidified Low-Level Wastes. The Literature Survey." Brookhaven National Laboratories (BNL-51899).

31.   Spence, R. D., W. D. Bostick, E. W. McDaniel, T. M. Gilliam, J. L. Shoemaker, O. K. Tallent, I. L. Morgan, B. S. Evans-Brown, and K. E. Dodson. 1989. "Immobilization of Technetium in Blast Furnace Slag Grouts." *Fly Ash, Silica Fume, Slag, and Natural Pozzolans in Concrete, Proc. 3rd Int. Conf.* American Concrete Institute, Detroit.

32.   Kharaka, Y. K., W. D. Gunter, P. K. Aggarwal, E. H. Perkins, and J. D. DeBraal. 1988. "A Computer Program for Geochemical Modeling of Water-Rock Interaction." U.S. Geological Survey, Water-Resources Investigations Report 88-4227.

33.   Hegelson, H. C. 1969. "Thermodynamics of Hydrothermal Systems at Elevated Temperatures and Pressures." *Am. J. Sci.* 267:729–804.

34.   Marshall, W. L. 1980. "Amorphous Silica Solubility. III. Activity Coefficient Relationships and Prediction of Solubility Behavior in Salt Solutions, 0–350°C." *Geochim. Cosmochim. Acta* 44: 925–931.

35.   Andersson, K., B. Allard, M. Bengtsson, and B. Magnusson. 1989. "Chemical Composition of Cement Pore Solutions." *Cement Concrete Res.* 19: 327–321.

INDEX